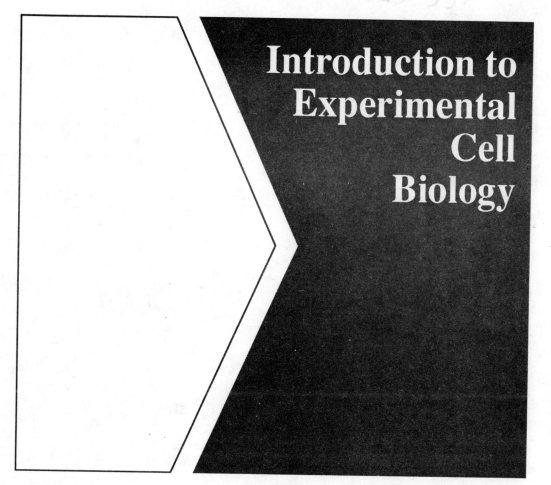

Introduction to Experimental Cell Biology

Introduction to Experimental Cell Biology

Holly Ahern

State University of New York, Albany

Wm. C. Brown Publishers

Book Team

Editor *Kevin Kane*
Developmental Editor *Carol Mills*
Production Coordinator *Kay Driscoll*
Photo Editor *Robin Storm*
Permissions Editor *Karen L. Storlie*

 Wm. C. Brown Publishers

President *G. Franklin Lewis*
Vice President, Publisher *George Wm. Bergquist*
Vice President, Operations and Production *Beverly Kolz*
National Sales Manager *Virginia S. Moffat*
Group Sales Manager *Vincent R. Di Blasi*
Vice President, Editor in Chief *Edward G. Jaffe*
Executive Editor *Earl McPeek*
Marketing Manager *Paul Ducham*
Advertising Manager *Amy Schmitz*
Managing Editor, Production *Colleen A. Yonda*
Manager of Visuals and Design *Faye M. Schilling*
Production Editorial Manager *Julie A. Kennedy*
Production Editorial Manager *Ann Fuerste*
Production Editorial Manager *Vickie Putman Caughron*
Publishing Services Manager *Karen J. Slaght*

WCB Group

President and Chief Executive Officer *Mark C. Falb*
Chairman of the Board *Wm. C. Brown*

Cover design by Morris Lundin
Cover photo courtesy of Joseph De Pasquale, State University of New York at Albany
Copyedited by Denise Lightle

Printed in the United States of America by Wm. C. Brown Publishers, 2460 Kerper Boulevard, Dubuque, IA 52001

10 9 8 7 6 5 4 3 2

Contents

Appendix

Preface

Cell biology encompasses a wide range of scientific disciplines—from the simple microscopic study of cells or investigations into their molecular basis to studies on the cellular organization of highly complex multicellular organisms. But how does one go about designing a laboratory course for a subject that is so diverse and far-reaching? The approach taken in this manual is to provide a comprehensive overview of cell biology, starting with the basics (microscopy) and working up to the more complex (animal cell culture). Along the way, students are taken inside the cell to study the important macromolecules that run it (nucleic acids and proteins), and peek at the complicated machinery that is essential to normal cell functions.

This book is primarily directed at junior and senior level undergraduate biology majors or first-year graduate students who have a basic knowledge of cell biology and biochemistry. It was written not only to introduce students to the experimental side of cell biology but to present them with an idea of what "real science" is all about. Students can't be expected to learn cell biology by just hearing about it in lectures or going to a lab and performing boring manipulations on dead or prepared material. They must be exposed to the living science that cell biology really is. The experiments in this manual are designed to promote scientific thinking and independent thought in addition to illustrating concepts in cell biology and to help students learn underutilized yet essential scientific skills, especially the art of scientific writing. Perhaps the most important thing to note about this manual is that it works. It has been extensively tested by the greatest critics of all—my students.

In most real life research situations, experiments are started on one day and then continue for several days before the final results can be determined. However, students at colleges and universities may find it difficult to fit one or two hours of lab time every day into their busy schedules. The exercises in this manual were developed to fit into a weekly four-hour lab period without losing the feel for what really goes on in research laboratories. They are flexible enough to fit other lab formats as well. On some occasions, students will have to return to the lab on subsequent days, but only for a short time—to make observations or to pick up results.

The cell culture experiment is an exception to the rule of having lab beyond the regular four-hour period. Because animal cells in culture will die if they are not periodically subcultured, students will have to return to the lab daily to examine their cultures, and every fourth day to subculture during the two-week duration of this experiment. However, giving the students the responsibility of growing and maintaining cells and then finally using those carefully kept cells in an experiment keeps them actively involved and interested in the whole process. Every year the cell culture experiment earns the highest marks from my students.

The manual is designed to be user-friendly. Each experiment starts off with a descriptive introduction, followed by detailed guidelines for instructors to ensure the success of each experiment. For the most part, the biological materials used in the experiments should be readily available in a well supplied biology department at any college or university. Recipes for media and reagents can be found in the appendices, along with additional procedures that instructors may find useful.

This laboratory manual was written with one goal in mind: to provide students with a comprehensive and meaningful laboratory experience in experimental cell biology. Hopefully, upon completion of this course, students will be inspired to continue in the field after graduation. Even if they do not, it will provide them with some of the skills necessary to get a job or go on to other studies—in medicine or in graduate school, for example.

I owe a special note of thanks to a number of people who were involved in the development of th's book. My enthusiastic and hardworking students deserve much of the credit, as their suggestions and comments on how to improve the experiments were invaluable. The faculty of the cell and molecular biology core areas at the State Univeristy of New York at Albany provided me with resources and technical support throughout the long gestation of this book. Finally, very special thanks to my family, Kevin, Marty, and Kaleigh, for their love and support, from the beginning of this project to the end.

Format for Laboratory Reports

Laboratory reports will take the format of a scientific paper but in condensed form. A scientific paper, such as that submitted to scientific journals for publication, has six sections: Abstract, Introduction, Materials and Methods, Results, Discussion, and References. Laboratory reports for this course will consist of four of these sections: the Introduction, Results, Discussion, and Reference sections. It is recommended that students write at least one full-length scientific paper on a topic of their choice to gain experience in this understressed but very important laboratory skill.

Laboratory reports must be typed, and at least three references should be used in the preparation of the paper. These references can include textbooks, reference manuals, journal articles, or material obtained in a lecture or interview. It is also a good idea to go to the library and read a few current research articles from a well-known scientific journal (the journal *Cell* is an excellent choice), simply to get a feel for the format and style used in writing a scientific paper.

The format for laboratory reports is given below.

Title Page

Use an inciteful title, not just the name of the experiment copied out of the lab manual. Pick a title that describes the contents of your paper. For example, for a paper on microscopy:

Bad title—The Microscope in Cell Biology
Better title—A Comparison of Various Microscopy Techniques Used to Examine Stained and Unstained Biological Material

Introduction

The introduction includes background information—on methods used, on previous research in the same area that is pertinent to your research, on historical aspects of the experiments, and on anything that needs to be explained so that a person who was unfamiliar with the topic could understand the importance of your research. The purpose of your research should be included in the Introduction, usually in a sentence or short paragraph at the end of the section.

Any nonoriginal information contained in this section (or any other section) must be cited (see section on references).

Results

The data obtained from your research is presented in this section, in a logical and unbiased manner. Data should not actually be interpreted in the Results section—just present what happened but not why (save the whys for the Discussion). A well-written Results section will include a paragraph or two that briefly explains how the data were obtained. The actual data should be presented in the form of appropriately labelled tables and/or figures. Numbers or other hard data should be shown in a table. Graphs, diagrams, photocopies, or pictures are shown as a figure (see example below).

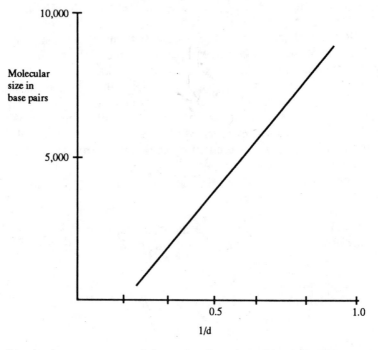

Figure 1. Standard curve generated from the digestion of lambda DNA with the restriction enzyme Hind III.

Table 1. Restriction fragment sizes determined from the digestion of the recombinant plasmid pMB1 with the restriction enzyme Pvu II.

Band	Distance from Well (cm)	1/d	Molecular Size (in base pairs)
1	3.2	0.31	4,500
2	5.0	0.20	2,000
3	5.6	0.18	1,750

x

Discussion

In the Discussion section, the data presented in the Results section is interpreted and discussed. The most important thing to remember about writing a discussion is not to reiterate the results! A well-written Discussion might include the following:

1. What were the major points illustrated by the data?
2. Do the results agree with previously published works? (the previously published works must be referenced)
3. Is the data contradictory in itself?
4. Does your research have potential for follow-up experiments?

State a conclusion at the end of this section. Your conclusion should *not* be ". . . the experiment was simple and easy and fun to do." If you cannot come up with a pertinent and logical conclusion, do not include one.

References

Any information obtained from a nonoriginal source must be referenced. Footnoting or endnoting is generally not an acceptable means of citing literature for scientific writing. The following is a widely used means of referencing other works.

At the end of the paper, include a section entitled "References" or "Literature Cited." Each piece of literature used in the preparation of your paper should be listed, alphabetized by first author, and numbered. For example—

1. Dorin, J. R. 1987. 'A clue to the basic defect in cystic fibrosis from cloning the CF antigen gene.' *Nature* 326:614-617.
2. Prescott, D. 1988. **Cells**. Jones and Bartlett Publishers.

Wherever necessary in the text of your paper, cite the references as shown below:

This result agrees with the data obtained by Dorin (1) . . . Fibronectin has been found on the surface of fibroblasts grown in culture (2).

Remember, when in doubt, cite it! It is better to reference too much than not enough.

Chapter 1

The Microscope in Cell Biology

Introduction

For well over a century, the light microscope has been one of the most important instruments available to biologists of all disciplines. In the past two or three decades, the light microscope has been modified and improved in many ways. Various contrast generating devices were added to improve image quality, and these developments led to the invention of the phase-contrast microscope, differential-interference-contrast (DIC) microscopes, the electron microscope, and the technique of X-ray diffraction.

The bright-field, or light, microscope is the starting point for a discussion of modern cell biology. Before its invention, scientists were unable to view individual cells but were instead confined to gross examinations of whole tissues, a practice that supplied very little information about the composition or function of cells. The basic function of the bright-field microscope, shown in figure 1.1, is to magnify or enlarge an object so that it becomes visible to the observer. It can be used to view amplitude objects, which are visible because they absorb light.

Magnification of an amplitude object is achieved using a two-lens system, composed of the ocular lens and the objective lens(es). Most bright-field microscopes in use today are equipped with four objective lenses, with magnifying powers of $4\times$ (scanning), $10\times$ (low power), $45\times$ (high power), and $100\times$ (oil immersion). The objective lens is closer to the specimen and magnifies it, producing the "real image". The real image is then projected up the body tube to the oculars, which further magnify the specimen by $10\times$ to produce the final image. The total magnification achieved is determined by multiplying the magnifying power of the objective and ocular lenses, as shown below:

Ocular Lens	Objective Lens	Total Magnification
$10\times$	$4\times$	$40\times$
$10\times$	$10\times$	$100\times$
$10\times$	$45\times$	$450\times$
$10\times$	$100\times$	$1000\times$

It is not possible to achieve unlimited magnification in a light microscope because lenses are limited in their ability to magnify by a property known as resolving power. Resolving power is the ability of a lens to show two separate objects as discrete entities.

1

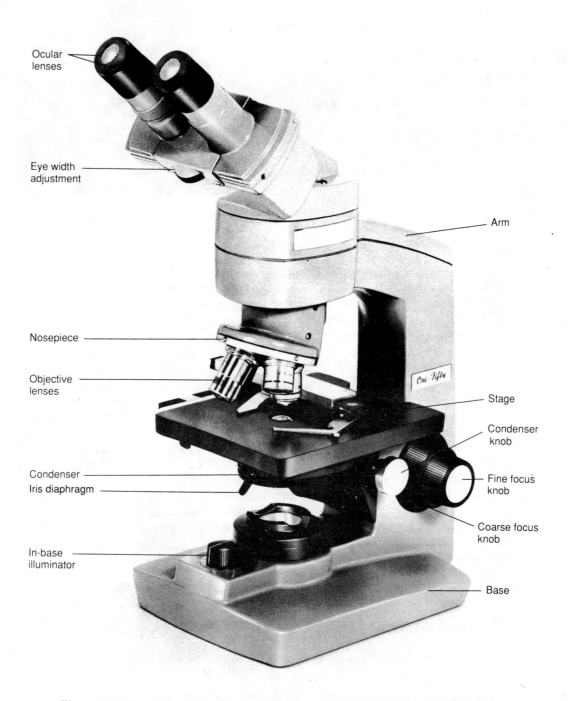

Ocular
lenses

Eye width
adjustment

Nosepiece

Objective
lenses

Condenser

Iris diaphragm

In-base
illuminator

Arm

Stage

Condenser
knob

Fine focus
knob

Coarse focus
knob

Base

Figure 1.1. The bright-field microscope. (Courtesy of Reichert Scientific Instruments)

When a lens becomes unable to discriminate (that is, when two objects appear as one), it is said to have lost resolution. Increasing the magnification will only decrease the quality of the image. The resolving power of a lens (d) can be expressed in the following formula, where d is the minimum distance between two discrete entities, α is the aperture angle of the lens, and n is the refractive index of the surrounding medium:

$$d = \frac{0.6\,\lambda}{n\,\sin\alpha}$$

From this equation it can be seen that the shorter the wavelength of light, the greater the resolving power of the lens. This equation also shows the relationship between refractive index and resolution. The refractive index (n) is the bending power of light as it passes through a medium to the objective lens. The refractive index can be defined as the ratio of the velocity of light *in vacuo* to that in the medium:

$$n = \frac{v(vac)}{v(med)}$$

By increasing the refractive index of the media, decreased light refraction occurs and more light rays enter directly into the objective lens. This explains why immersion oil must be used when working with the $100\times$ objective lens. Immersion oil has the same refractive index as glass; therefore, as light passes through the slide, then through the oil, only minimal light refraction occurs. A vivid image with high resolution is the net result.

In order for amplitude objects to be visible at all, they must first be fixed permanently to a slide and then stained. Treatment of the specimen with heat or chemicals cross-links cellular proteins so that cells are stabilized and locked in position on the slide. The specimen can then be stained for visualization in a light microscope. There are many different staining techniques that can be used to stain entire cells, or specific components of the cell, such as the nucleus or Golgi apparatus. The process of fixation kills cells and causes distortions, therefore cells viewed after fixation and staining are not necessarily seen as they exist in nature. Until the phase-contrast microscope became widely available after World War II, cell biologists obtained all of their information about cell structure ← from dead, stained material.

In 1932 Fritz Zernicke invented the phase-contrast microscope, giving biologists their first look at the activities of living cells. The phase-contrast microscope is basically a light microscope outfitted with a special kind of optical system designed to exploit the diffracting properties of cells to generate contrast and render unstained cells visible. The phase-contrast microscope is used to view phase objects, which are simply objects composed of material with a refractive index unlike the medium that surrounds them.

In phase-contrast microscopy, light must be considered as waves, rather than photons. Light is emitted from a source either at a characteristic frequency (monochromatic light), or at a series of frequencies (polychromatic light). The frequency (f) is invariant, but as the light travels through various media of different refractive index, for example air, water, or cytoplasm, its velocity (v) and wavelength (λ) vary reciprocally:

$$f = \frac{v}{\lambda}$$

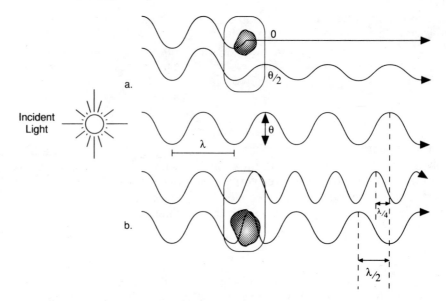

Figure 1.2. Light waves passing through a cell: (a) a fixed and stained cell (amplitude object) (b) a living, unstained cell (phase object).

As light waves enter a medium from the air, they are slowed down, or retarded. On re-emerging they assume their original velocity. As the light waves pass through a phase object, they are shifted in phase (or phase shifted) relative to those passing near the object through the medium, as shown in figure 1.2. As light passes through a living cell, the phase of the light waves is changed. Light passing through a thick or dense part of the cell, such as the nucleus, is retarded and its phase consequently shifted, relative to light passing through an adjacent thinner region, such as the cytoplasm. Contrast is therefore achieved by the constructive and destructive interference of the light waves when they are recombined by the microscope to form an image of the cell.

In order to achieve phase-contrast, two special parts are added to a bright-field microscope. The first is a phase annulus, a fully silvered or black painted plate with an annular (ring-shaped) window. The phase annulus replaces the iris diaphram in the condenser, the lens which focuses the incident light. Light passes through the condenser from the annulus as a hollow cone of light coming to a point at the specimen plane. This light then diverges after the specimen plane into an inverted hollow cone of the same angle, which passes through the objective lens, as shown in figure 1.3.

The second component is the phase plate, which is usually a curved lens surface with a ring-shaped area coated with a light-absorbing material. The purpose of the phase ring is to advance or retard light waves by ¼λ. This results in the diffracted and undiffracted light waves showing destructive interference, which is what creates contrast in a phase-contrast microscope.

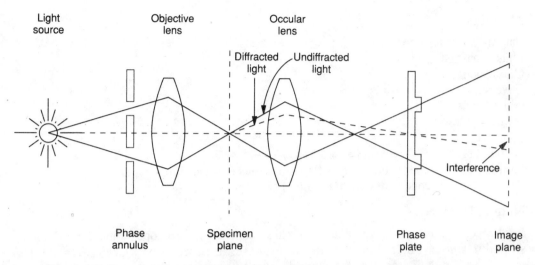

Figure 1.3. Ray diagram showing the optics of a phase-contrast microscope.

To maximize contrast, two things should be done. The first is to adjust the microscope for Kohler illumination. Kohler illumination is used to obtain uniform illumination of a field to get the best resolution, contrast, and image quality possible. The steps required to achieve Kohler illumination are:

1. Using the low power objective, close the iris diaphragm almost completely.
2. Focus the specimen plane (it is helpful to have a specimen in the plane).
3. Close the field diaphragm and adjust the condenser so as to focus the edge of the field diaphragm at the specimen plane.
4. Open the field diaphragm and adjust the iris diaphragm.

After Kohler illumination is obtained, the phase-contrast microscope must be further adjusted by centering the phase annulus. The phase annulus is centered using screws in the condenser mount so that the bright annulus is superimposed over the dark ring of the phase plate. This adjustment is usually made while looking down the empty microscope tube (with the ocular lens removed) or while looking through a phase telescope or Bertrand lens. It is worthwhile to look at a living cell or other specimen while centering and decentering the phase annulus to see how the image behaves when adjustment is correct. As the annulus moves over the phase plate, the field darkens considerably and the object takes on a much higher contrast.

Many other types of microscopes are used in cell biology. These include the dark-field microscope, differential interference microscope, fluorescence microscope, and the electron microscope, which utilize various optical systems and stains to generate contrast in order to view the intricate details of cells. This helps scientists to better understand cell processes such as chemotaxis (cell movement) and cell division. New staining techniques, such as immunostaining using antibodies specific for cell components, have also increased the power of the microscope as a research tool.

6

Materials

Bright-field microscope
Phase-contrast microscope
Glass slides
Coverslips
Glass beads
Immersion oil
Toothpicks
Trypan blue (0.4% Trypan blue in dH_2O)
Humidified chamber (petri dish with a small piece of wet filter paper)
Prepared microscope slides with a stained specimen

Note to Instructors

Proper use of the bright-field and phase-contrast microscopes should be demonstrated to students before they begin this exercise. If members of your department use microscopes in their research (for example, an electron microscope), try to arrange demonstrations for the students to provide them with an overview of the various methods used in cell biology.

Procedure

1. Place a slide with a stained specimen on the stage of the phase-contrast microscope and focus the image without adjusting the illumination. Note the appearance of the specimen. Then, using the directions from the Introduction, adjust the microscope for Kohler illumination and examine the specimen again. Compare the two images.
2. Using the directions from the Introduction, center the phase annulus on the phase-contrast microscope with a specimen on the stage. Note the changes in the image as the annulus is superimposed over the phase plate.
3. Clean and label three glass slides. Place a few glass spheres on each slide and cover them with a coverslip. Add water (n = 1.3330) to one, immersion oil (n = 1.5150) to the second, and leave the third dry (n of air = 1.0000). View the preparations in order of increasing refractive index. Experiment with the illumination (field diaphragm and condenser) to control contrast. Note how changes in the refractive index effects the image of the spheres.
4. Make two "spit cell" preparations on labelled slides. Add a small drop of water to each slide. Using a toothpick, gently scrape off a little cheek tissue (which contains squamous epithelial cells) and mount it in the drop on the slide. Cover the preparation with a coverslip. Place one of the slides in a humidified chamber for later examination. Examine the other slide using the bright-field microscope. Remember to regulate the illumination to get good contrast. Note the bright-field image, looking closely at the cell surface.

   Examine the second slide using the phase-contrast microscope, making the same observations as above. Compare the images seen.

5. Make a stained spit cell preparation on a labelled slide. Place a drop of Trypan blue on a clean slide, then mount some cheek cells in the drop on the slide. Gently stir the two together. Apply a coverslip and examine individual cells using high power with the light microscope.

6. Compare the image of the stained and unstained cells seen with the bright-field microscope. Compare the stained cells to the unstained cells viewed by phase-contrast microscopy. On a drawing of an individual stained cell, label the following components:

   Nucleus (spherical darkly stained body near the center of the cell)

   Chromatin (stained threadlike material within the nucleus)

   Nucleolus (tiny darker stained body within the nucleus)

   Cytoplasm (lightly stained fluid within the cell membrane)

   Cytoplasmic organelles and inclusions (small, sometimes darkly stained bodies found in the cytoplasm)

   Plasma membrane (boundary of the cell)

Chapter 2

Introduction

In the late nineteenth century, Gregor Mendel postulated that the genetic units involved in inheritance were physically transmitted from parent to offspring. However, it was not determined until many years later that the units actually involved in inheritance of genetic traits were genes, carried on the chromosomes within the nucleus of all cells. After further speculation and research, genes were found to be molecules of deoxyribonucleic acid (DNA). However, the exact role of DNA in heredity was a mystery until 1953, when James Watson and Francis Crick proposed that the three-dimensional structure of DNA was a double helix; two complementary polynucleotide strands wound around one another. Soon after it was proposed that inheritance of traits was actually due to separation of the polynucleotide strands and synthesis of an identical strand using parental DNA as a template. This exact copy of the parental DNA was then passed on to offspring following cell division.

Since the discovery of the double helix, genes have become the subject of intense research. A full understanding of the structure and function of genes will lead to greater insights into the molecular basis of heredity and could also lead to treatments and perhaps cures for genetic diseases and some forms of cancer.

It is difficult to study the organization of the genetic material because cells are usually in interphase, a period when the DNA does not replicate. The chromosomes are diffuse and tangled within the nucleus during this period. Individual chromosomes are only distinguishable during meiosis or mitosis, when the chromatin coils up into a compact structure that can be viewed microscopically; however, the study of mitotic chromosomes yields little information about their function.

However, in some cells from *Drosophila* and other fly larvae, the chromatin is organized into large individual chromosomes during interphase. These chromosomes are called polytene (meaning "many-threaded"). They are formed by repeated replication of the DNA in the chromosomes without cell division, resulting in the formation of thousands of identical parallel helices. Polytene chromosomes can be clearly seen using a light microscope because they are large and remain in a long, stretched-out configuration as opposed to the coiled form in mitotic chromosomes. They are most often studied in the salivary gland cells of *Drosophila* but are found in other secretory cells and some ciliated protozoa and plants as well.

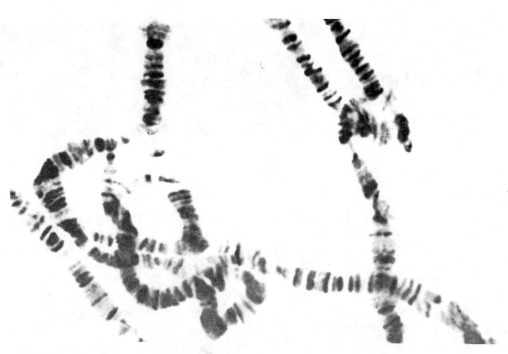

Figure 2.1. Polytene chromosomes from *Drosophila* as seen by light microscopy. David G. Futch "A study of Speciation in South Pacific Populations of Drosophilia ananassae," in Marshall R. Wheeler, ed. Studies in Genetics, no. 6615 (Austin: University of Texas Press 1966).

Polytene chromosomes have been studied for many years in an effort to determine the organization of genetic material within cells and to learn more about gene expression. When viewed with a light microscope, polytene chromosomes show a distinctive pattern of alternating dark and light interband regions (figure 2.1). This pattern can vary between different chromosomes. Studies on mutant flies suggest that each band contains only a single gene that is represented once in each of the strands that make up the chromosome. During gene expression, the DNA within a particular band will begin to unfold, and the gene contained within that region is expressed. Gene expression is evidenced by the incorporation of radioactive nucleotide precursors into these regions during periods of gene activity. RNA has also been directly isolated from these regions, called "puffs" because of their puffed appearance, indicating that puffed bands represent sites of intense RNA synthesis. Puffing occurs on polytene chromosomes in a sequential pattern, which corresponds to the ordered pattern of gene expression seen during *Drosophila* development. Polytene chromosomes, then, provide researchers with an excellent experimental system for the study of gene expression.

In this experiment, you will isolate polytene chromosomes from the salivary glands of *Drosophila* third instar larvae. You will stain the chromosomes with orcein stain and then study the location of bands and look for the presence of puffs or sites of gene expression.

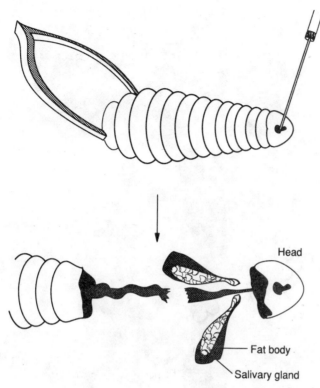

Figure 2.2. Method for dissection of fruit fly larva. (1) Grip the posterior end of the larvae with forceps. (2) Insert dissecting needle into mouth parts. (3) Steadily pull head away from the body, until larvae separates and salivary glands are pulled out onto the slide. (4) Remove the excess material (digestive tract, fat bodies, head). (5) Squash and stain the salivary glands.

Materials

Drosophila larvae, third instar stage
Dissecting needle
Forceps
Microscope slides
Coverslips
Dissecting microscope
Light microscope
Mayer fixative (Harleco)
0.7% saline
Orcein stain
70% ethanol in a squirt bottle
Petri dish with moist filter paper (incubation chamber)

Figure 2.3. Dissected *Drosophila* larvae. The salivary glands are labelled "S." C. Edward Gasque

Procedure

1. Clean two glass microscope slides with soap and water, and then with 70% ethanol. Allow them to completely air dry.
2. Place a small drop of Mayer fixative on one end of each slide. Using your finger, slowly wipe the fixative across the slide to form a thin layer over the entire surface. Put the slides aside to dry.
3. Using a dissecting microscope, remove the salivary gland from a *Drosophila* third instar larvae, as shown in figure 2.2.
   To aid in the recognition of these structures, note that each larvae will have two identical glands. They are very large, translucent, colorless and glistening. Each gland will have an associated darkly colored fat body, shown as "F" in figure 2.3.
4. Separate the fat bodies from both glands using a dissecting needle and forceps.

5. Scrape away as much excess material as possible from the glands on the slide. Add a few drops of 0.7% saline to wash the glands, then blot off the excess saline using the edge of a paper towel. Add a few drops of fresh saline to the slide and allow the glands to soak for 7 minutes.

6. Add a drop of orcein stain to the center of each of the two fixative coated slides.

7. Transfer one of the salivary glands to the drop of stain on one of the slides. Transfer the other gland to the second slide in a similar fashion.

8. Place the slides into a petri dish containing a moist filter paper. Incubate at room temperature for at least 15 minutes. If the stain begins to dry out during this time, add one or two drops more.

9. Remove one slide from the incubation chamber. Using the edge of a paper towel, blot away the excess stain, without touching the salivary gland.

10. Add one drop of fresh stain and allow the gland to soak for 1 minute.

11. Carefully place a coverslip directly over the gland. Place a paper towel over the slide so that it is centered over the coverslip. Press down with your thumb slowly and steadily in the very center of the coverslip, applying moderate pressure. Do not let the coverslip move from side to side during the squash process.

12. Repeat the squash with the second slide.

13. Check the squash preparation visually. A properly squashed gland will be two to three times its normal size.

14. View the preparation at 45× magnification using a bright-field microscope. Scan the slide for a cell with well stained nuclear material. A good prep will show long spread out chromosomes with distinct darkly stained bands. Scan the prep for the presence of puffs.

15. View the preparation at 100× (oil immersion). Identify the various features of polytene chromosomes including bands, interband regions, and puffs. Note the number and location of the puffed bands.

Chapter 3

Introduction

Genes, the units of heredity, are found on chromosomes within the nucleus of all living cells. The exact replication and distribution of the chromosomes at cell division are fundamental to heredity and, therefore, life. The actual process of cell division begins with the replication of each chromsome's DNA double helix. The daughter helices form two daughter chromosomes, which are then distributed to daughter cells at cell division. This process, known as the cell cycle (figure 3.1), is repeated with alternate rounds of DNA replication and distribution to daughter cells. In eukaryotic cells, with the exception of the gametes, the process by which chromosomes are distributed at cell division is called mitosis.

Mitosis consists of four stages: prophase, metaphase, anaphase, and telophase. A fifth stage, interphase, represents the time in between cycles of active cell division. During interphase, the cell doubles its mass and DNA replication occurs in preparation for cell division. In prophase, the duplicated chromosomes begin to coil up and condense. The condensed chromosomes then begin to move toward the central region of the cell and

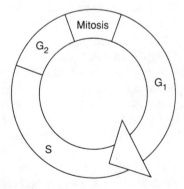

Figure 3.1. The cell cycle. The genetic material is replicated during the S period, and distributed to daughter cells during mitosis.

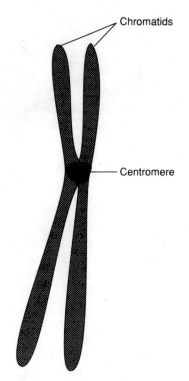

Figure 3.2. A metaphase chromosome; two chromatids connected at the centromere.

line up along a central axis during metaphase. In anaphase, the duplicated chromosome pairs begin to separate, and move toward opposite ends of the cell. Finally, in telophase, daughter nuclei begin to form as the chromosomes become decondensed and a nuclear envelope forms. At the end of telophase, the cell undergoes cytokinesis to form two daughter cells, each containing a full complement of chromosomes.

The analysis of the chromosomes in a eukaryotic cell that can provide researchers with a detailed picture of an organism's genetic makeup is referred to as karyotyping. Karotyping is performed on cells in metaphase. During metaphase, chromosomes in the nucleus have duplicated, and the duplicate chromosomes, called chromatids, are connected at the centromere (figure 3.2). The different metaphase chromosomes can be identified by their characteristic lengths and the location of their centromere. The centromere is classified according to its location on a metaphase chromosome as telocentric (centromere at one end of the chromatids), acrocentric (centromere close to one end), submetacentric (centromere near the middle), or metacentric (centromere in the middle) (figure 3.3). The centromere is always located at the same position for any particular chromosome, but the location differs among various chromosomes. For example, in a human cell, chromosome 1 is always metacentric, but differs from chromosome 13, which is always telocentric.

At any given time, the percentage of cells in mitosis, even in an exponential culture, is very small. The percentage of cells in metaphase is smaller yet, decreasing the likelihood of finding chromosomes suitable for karyotyping. In order to increase the per-

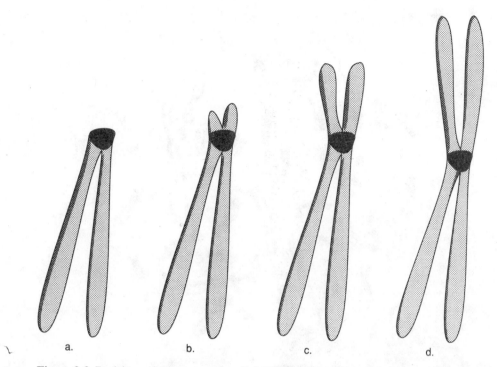

Figure 3.3. Positions of the centromere. (a) Telocentric (b) Acrocentric (c) Submetacentric (d) Metacentric.

centage of acceptable cells, they are first incubated with a mitogen such as phytohemagglutinin (PHA) to stimulate cell division. The frequency of metaphase cells can be further enhanced by adding a mitotic spindle inhibitor, such as colcemid, to exponentially growing cells. This arrests the cell cycle in metaphase by preventing spindle formation. In this manner, the number of metaphase cells in the culture will be substantially increased.

To prepare a karyotype from a culture of cells, the cells are broken open, spread on a glass slide, and then stained with special stains. The slide is then examined microscopically for the presence of metaphase chromosomes. In an actual karyotype, a series of metaphase preparations are photographed, and the individual chromosomes are cut out and arranged in order of decreasing size. This arrangement is called an idiogram. In most cases, groups of chromosomes are apparent, but it is not usually possible to identify individual chromosomes solely on the basis of size and position of the centromere. The pattern of bands produced on the chromosome by staining with dyes such as quinacrine (Q-bands) or Giemsa (G-bands) are characteristic for individual chromosomes and simplifies the process of chromosome identification. The four criteria generally used to define the karyotype of a cell are the number of chromosomes, the individual lengths, the location of the centromere, and the banding patterns of individual chromosomes when stained.

18

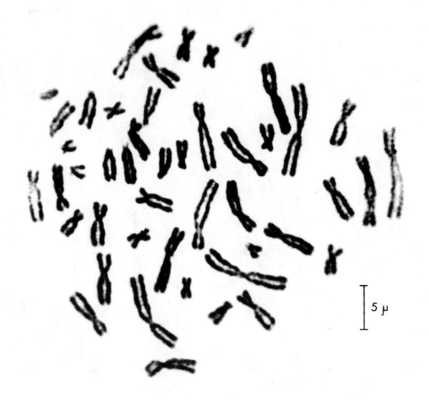

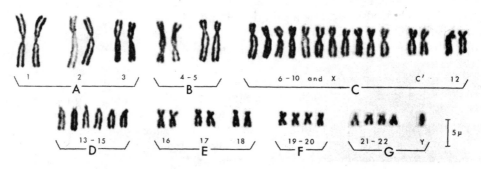

Figure 3.4. A chromosome karyotype from a human male. Note the X chromosome in group C, and the Y chromosome in group G. © Carolina Biological Supply Company

Normal human cells have 23 different chromosomes that exist within a cell in pairs called homologues. Chromosome pairs 1–22 are called autosomes and are numbered on the basis of decreasing size. The 23rd pair are the sex chromosomes, which carry genes determining the sex of an individual, as well as other genes. In females, the sex chromosomes are both X chromosomes (so called because they look like X's). In males, one of the sex chromosomes is an X, while one is a much smaller chromosome called a Y chromosome. An example of a normal human karyotype is shown in figure 3.4.

When performed properly, a human karyotype can provide invaluable genetic information. Certain diseases, such as Down's syndrome, can be diagnosed from karyotypes performed on fetal cells. In most clinical cytogenetics laboratories, lymphocytes from the blood are used to prepare karyotypes of human cells.

In this experiment, you will culture lymphocytes from human blood and prepare chromosome spreads from the cultured cells. You will stain the spreads for G-banding by partially digesting the chromosomes with the proteolytic enzyme trypsin in the presence of Giemsa stain. After the chromosomes have been banded, you can distinguish and identify them by their characteristic size and banding pattern.

Materials

Heparinized whole human blood
Chromosome Medium 1A (Gibco)
Colchecine (colcemid 10 μg/ml—Gibco)
0.075 M KCl
Absolute methanol
70% methanol (in water) at 4° C
Glacial acetic acid
Phosphate buffer (0.025 M phosphate buffer pH 6.8)
1% trypsin (in sterile, distilled water)
Stock Giemsa stain (0.4% w/v in methanol pH 6.9—Sigma)
37° C water bath
Clinical centrifuge
CO_2 incubator
15cc plastic culture tubes
Pasteur pipettes
Petri dishes
Glass slides
Forceps
Bunsen burner
Bright-field microscope
Immersion oil

Note to Instructors

Most of the materials required for this experiment can be purchased as a kit from Gibco (catalog number 120-6706AV). Only five drops of blood are needed for the initial lymphocyte culture, which can be obtained via individual fingersticks or by drawing one tube of blood for the class. The student or instructor can start the cell cultures up to 4 days before the scheduled lab period.

Procedure

Two to four days prior to lab:

1. Add 5 mls of Chromosome Medium 1A to a 15cc plastic culture tube (this medium contains the mitogen phytahemagglutinin—PHA). Place a tube in a 37° C water-bath and allow the media to prewarm for 10 minutes.
2. Add 5 drops of whole blood to the prewarmed media.
3. Incubate the tube with the cap loosened in 5% CO_2 for 48–96 hours at 37.5° C. Mix the contents of the tube by inversion at least once daily (remember to tighten the cap before inverting the tube).

Two to four hours before lab:

4. Add 160 μl of colcemid (final concentration 0.4 μg/ml) to the tube and mix by inversion. Colcemid is a spindle inhibitor and blocks the cells in the metaphase stage of their cell cycle. The addition of colcemid synchronizes the cells and maximizes the number in metaphase before the preparation of the chromosome spreads.
5. Return the tubes to the incubator until the lab period begins.

start ⟶ **Preparation of the Chromosome Spread**

6. After at least 2 hours of incubation with colcemid, centrifuge the cells at 400 × g for 8 minutes. Using a pasteur pipette, carefully remove and discard the supernatant, leaving approximately 0.5 ml of liquid on the cell pellet. *1200 RPM*
7. Tap the tube with your fingers once or twice to resuspend the cells. Add 6 ml of hypotonic solution (0.075 M KC1) 2 ml at a time to the cell suspension, mixing gently after each addition.
8. Incubate the tube in a 37° C water bath for 10 minutes. During this incubation, prepare fixative solution by mixing 4 ml of acetic acid with 12 ml of methanol. Place the fixative solution on ice. *2* *'burn use 6 pipette*
9. Centrifuge the cells at 400 × g for 8 minutes at room temperature. Using a pasteur pipette, carefully aspirate the supernatant and discard. The cells are swollen and very fragile in this state; handle them gently.
10. Add 6 ml cold fixative solution to the cell pellet, 2 ml at a time with gentle mixing, and place the tube on ice.
11. Incubate the tube on ice for 20 minutes.
12. After the incubation, centrifuge at 400 × g for 10 minutes, then aspirate the supernatant and discard.
13. Add 6 ml of fresh fixative, 2 ml at a time with gentle mixing, and incubate the tube on ice for 20 minutes.
14. Centrifuge the tube at 400 × g for 10 minutes. Discard the supernatant, leaving approximately 0.5 ml of fixative on the pellet. At this point, the pellet contains mostly white blood cells and may be difficult to see.
15. Gently resuspend the pellet in the remaining fixative, using only the thin bottom part of a pasteur pipette. You can store the cells in this state for up to a day at 4° C or use them immediately to make a chromosome spread.

supernate
½ ml of liq.
on u porion use pasteur pipet
Blood - poor in Bucla

mix good
add fix 2mls at time

```
  218          327
   10           20
  228        → 347
  230
    8         346
  238          18
  247
   20          356
  307
  110
  317
```

16. Dip several alcohol-cleaned slides in a solution of ice-cold 70% methanol. Allow the excess liquid to drip from the slide, but do not dry it off.

17. Prop three or four slides so that they form a 45° angle with the bench top. Using a pasteur pipette held at least 2 feet above the slides, drop several drops of cell suspension down onto the slides below (6–8 drops per slide is usually sufficient). Try to place the drops toward the top of the slides so that the cell solution drips downward and spreads over the surface of the slide.

18. ~~Pass the slides briefly through the flame of a bunsen burner~~. Place the slides in a 37° C incubator until they are completely dry. At this point the slides can be stained or stored for up to a week at room temperature before staining.

Staining for G-band Detection

19. Prewarm phosphate buffer (0.025 M phosphate buffer, pH 6.8) for 10 minutes in a 60° C water bath.

20. Incubate one of the dried slides in the prewarmed phosphate buffer for 10 minutes.

21. Prepare 10 ml of working Giemsa stain by adding:

    6.25 ml phosphate buffer
    0.25 ml 1% trypsin solution
    2.5 ml methanol
    1 ml stock Giemsa stain (**toxic:** wear gloves) to a plastic culture tube. Close the cap and mix by inversion.

22. Place the treated slide in a petri dish on a level surface. Flood the surface with working Giemsa stain and incubate for 20 minutes at room temperature.

23. Using forceps, remove the slides from the petri dish and rinse with tap water. Allow the slides to air dry completely, then examine briefly using the 45× objective using a bright field microscope. If the stain does not appear to be adequate, repeat the staining process on the other slides, increasing the incubation time.

24. Examine the well-stained slides using the 100× objective (oil immersion). Look for cells whose contents have been uniformly spread on the surface of the slide. Examine the nuclear material. Look at the overall appearance of the chromosome spread. Try to identify some of the chromosomes by determining the relative size and position of the centromere. Compare them to chromosomes from an actual idiogram (figure 3.4). Look especially for the X and Y chromosomes.

Chapter 4

Introduction

Plasmids are small, circular molecules of DNA found in bacterial cells. These extrachromosomal bits of double stranded DNA are found in a wide range of bacterial species and can be very small (fewer than 2000 base pairs in length) or very large (over 100,000 base pairs). Plasmids carry genes that are not absolutely required for host cell function, such as genes for resistance to antibiotics. Bacterial plasmids also carry genes that allow them to autonomously replicate within a bacterial cell.

When a single plasmid enters a susceptible host cell, it replicates until its characteristic copy number is reached. The copy number, or number of plasmids per bacterical cell, is dependent upon the genetic makeup of both the plasmid and host cell and ranges from one copy up to hundreds of copies per cell. Once the characteristic copy number is reached, replication of the plasmid DNA is shut off, and the plasmids are maintained within the cell as an extra genetic unit. When a bacterial cell containing a plasmid divides, the plasmids are also inherited so that each daughter cell receives at least one plasmid. In this way, plasmid molecules are propagated in the bacterial population.

In recent years, plasmids have become the workhorses of scientists from many disciplines. These incredible molecules are easily extracted from the host bacterial cells and can be manipulated genetically. Using recombinant DNA techniques, genes of interest from any source, a human gene for example, can be inserted into a plasmid. The recombinant plasmid is then transformed into host bacterial cells. Once inside the bacteria, the gene can be induced to synthesize its encoded protein, which can then be purified and studied biochemically. Or the gene can be sequenced to determine its nucleotide composition, or it can simply be stored for later use. In this way, copies of important genes are saved for future studies.

In order to utilize the genes stored on plasmids, it is often necessary to remove the plasmid from its bacterial host cell. Isolation of plasmid DNA from host bacteria requires that it be separated from the cellular DNA and RNA, as well as other cellular macromolecules such as proteins. Plasmid DNA can be isolated from host cells by first lysing the cells and then selectively precipitating the contaminating macromolecules out of solution until all that remains is the plasmid DNA. Over the years a number of different methods have been developed to purify plasmid DNA. The most commonly used small-scale method was developed by Birnboim and Doly in 1979 and is called the alkaline lysis, or mini-prep, technique.

To isolate plasmid DNA, the host cells are first treated with sodium dodecyl sulfate (SDS) and lysozyme in an alkaline environment. This weakens the cell wall and lyses the cells. The lysate is then neutralized with acidic sodium acetate. This treatment selectively denatures high molecular weight (chromosomal) DNA without damaging the covalently closed circular (plasmid) DNA. The presence of salt in the neutralizing buffer causes the chromosomal DNA to aggregate into an insoluble network that precipitates out of solution. The high concentration of sodium acetate in the solution also precipitates most of the cellular RNA and the protein-SDS complexes that have formed as well. The precipitated contaminating macromolecules are pelleted out of solution by centrifugation, leaving plasmid DNA and some residual low molecular weight RNA in solution. The remaining nucleic acids can then be precipitated from the supernatant by ethanol or isopropanol precipitation. The plasmid DNA can then be completely purified from residual RNA by incubating the solution with the enzyme RNAse, which selectively degrades RNA but leaves the DNA intact.

When this techique is performed carefully, the yield of purified plasmid DNA is usually quite good. One way in which to estimate the amount of DNA isolated from a mini-prep is to use agarose gel electrophoresis. Heated molten agarose, when cooled, solidifies to form a solid matrix. Samples of DNA are placed in wells cut into a horizontal

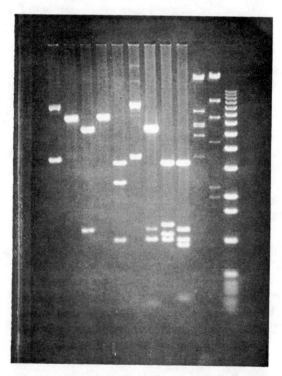

Figure 4.1. Ethidium bromide stained DNA in an agarose gel.

slab of agarose and electrophoresed. DNA molecules are negatively charged and will migrate toward the anode in the applied electrical field. Therefore DNA molecules can be separated in an agarose gel on the basis of size. Larger molecules migrate slowly through the matrix, while smaller molecules slither through the agarose meshwork and travel further toward the anode. The gel is stained with ethidium bromide, a fluorescent agent that intercalates between the bases of DNA. When the gel is placed on an ultraviolet light source, the DNA becomes visible as light bands against the dark background of the gel (figure 4.1).

In this experiment, you will isolate the plasmid pRPC245 by performing mini-preps on transformed *Escherichia coli* (*E. coli*) cells, and estimate the yield of DNA by agarose gel electrophoresis. The DNA obtained from this mini-prep will be stored 4° C for use in a following experiment.

Materials

Solution I (50 mM glucose, 25 mM Tris-HCl pH 8.0, 10 mM EDTA, with 4 mg/ml lysozyme added immediately before use)
Solution II (0.2 N NaOH, 1% SDS, freshly made)
Solution III (5 M potassium acetate, pH 4.8)
Phenol:chloroform:isoamyl alcohol 25:24:1 (pre-mixed—Amresco)
TE buffer (10 mM Tris-Cl, 1 mM EDTA, pH 8.0)
1 μg/μl purified pBR322 DNA, stock solution (Sigma)
50× TAE
95% ethanol
Agarose (electrophoresis grade)
Ethidium bromide (10 mg/ml in water—Sigma) **Caution:** mutagen. Wear gloves when working with this reagent.
6× loading buffer
Microfuge tubes
Microcentrifuge
Ice
Micropipettor and sterile tips
Agarose Mini-gel apparatus (Bio-Rad)
Power supply
Sterile distilled water
E. coli MM294/pRPC245 overnight culture (see appendix II)

Note to Instructors

Twenty-four hours before the scheduled lab period, start a culture of *E. coli* MM294 transformed with pRPC245 in TY+Amp medium (2 ml per student), and grow it overnight at 37° C (see appendix III for instructions). If other strains with plasmids are available in your department, they can easily be substituted for the one described here.

The mini-gel apparatuses can be shared. Have students load their samples on one or two gels, and prepare one set of concentration standards for the class.

Ethidium bromide is a mutagen. Gloves should be worn when using this chemical.

Procedure

A. Agarose Mini-gel Preparation

Prior to or during the mini-prep procedure, prepare a 0.7% agarose mini-gel and 250 ml of running buffer as follows:

1. In a 200 ml flask, mix 0.35 g agarose with 52 ml of distilled water. Heat on a hot plate or in a microwave until the solution boils to dissolve the agarose.
2. Cool the solution slightly and add 1 ml of 50× TAE. Carefully add 2.5 μl of ethidium bromide and gently swirl to mix.
3. When the molten agarose has cooled to approximately 60° C (touchable), pour it into the casting tray of a mini-gel electrophoresis apparatus. Put the comb in place and allow the gel to solidify.
4. While the gel is solidifying, prepare 250 ml of running buffer (1× TAE). Add 5 ml of 50× TAE to 245 ml of distilled water. Add 12.5 μl of ethidium bromide and swirl gently to mix.
5. When the gel has hardened, carefully remove the comb to form sample wells in the gel. Place the casting tray with gel into the electrophoresis apparatus.
6. Add running buffer to the buffer chambers until the surface of the gel is submerged.

B. Mini-prep Procedure

1. Fill a microfuge tube nearly full with an overnight *E. coli*/pRPC245 culture (approximately 1.5 ml per tube).
2. Centrifuge the tube for one minute at high speed in a microcentrifuge.
3. Without disrupting the cell pellet, remove as much medium as possible with a pasteur pipette.
4. Add 100 μl of ice-cold Solution I to the pellet. Vortex to completely resuspend the cells.
5. Incubate the tube at room temperature for exactly 5 minutes.
6. Add 200 μl of freshly made, room temperature Solution II. Close the cap and mix vigorously by rapidly inverting the tube 10 times.
7. Incubate the tube on ice for exactly 5 minutes.
8. Add 150 μl of ice-cold Solution III. Close the cap and gently mix by slowly inverting the tube ten times.
9. Incubate the tube on ice for exactly 5 minutes. The solution will begin to take on a milky appearance as proteins and high molecular weight DNA precipitate out of solution.
10. Centrifuge the tube for 5 minutes on high speed in a 4° C microcentrifuge to pellet the precipitate to the bottom of the tubes.
11. Using a micropipettor set at 400 μl, carefully remove the supernatant, which contains the DNA, and transfer it to a new microfuge tube. Discard the tube containing the protein-SDS-chromosomal DNA precipitate.

12. Add 400 μl of a phenol/chloroform/isoamyl alcohol mixture to the tube and mix by vortexing. Spin on high speed for 1 minute. Remove the upper aqueous phase, which contains the DNA, and transfer it to a new tube.
13. Repeat the extraction with an additional 400 μl of phenol/chloroform/isoamyl alcohol. Remove the upper aqueous phase and transfer it to a new tube.
14. Measure the volume of the DNA solution. Add 2 volumes of room temperature 95% ethanol to the tube, and mix by vortexing.
15. Incubate the tube at room temperature for 2 minutes. DNA is insoluble in ethanol and will precipitate out of solution.
16. Spin for 5 minutes in a room temperature microcentrifuge at high speed.
17. Decant the supernatant and invert the tube over a paper towel to drain off the ethanol.
18. Wash the pellets with 1 ml of 70% ethanol. Flick the tube with your finger several times, then recentrifuge.
19. Decant the supernatant and invert the tube over a paper towel to drain off the majority of the ethanol.
20. Evaporate any remaining ethanol by directing a stream of hot air from a blow dryer over the top of the open tube. Alternately, the tube can be left inverted with the cap open until the ethanol has completely evaporated.
21. Add 25 μl of TE buffer to the tube and allow it to stand at room temperature for 5 minutes. Resuspend the DNA by gently pipetting up and down with a micropipettor. DNA is stable in solution and can be stored long-term at either 4° C or at −20° C.

C. Estimation of DNA Concentration by Agarose Gel Electrophoresis

1. Prepare DNA concentration standards using 1 μg/μl pBR322 DNA, to 100 ng, 10 ng and 1 ng, as follows:
   a. Dilute the 1 μg/μl stock pBR322 to 100 ng, 10 ng and 1 ng per μl by performing 10-fold serial dilutions in sterile water (1 μl stock DNA into 9 μl sterile water, etc.).
   b. Remove 1 μl from the DNA stock and each dilution tube into labelled microfuge tubes.
   c. Add 9 μl sterile water and 2 μl 6× loading buffer to each tube and mix. The loading buffer contains two dyes, xylene cyanol and bromophenol blue, in a buffered glycerol solution. These dyes do not actually stain the DNA but serve to monitor the extent of migration during electrophoresis.
2. Add 2 μl of mini-prep DNA to 8 μl sterile distilled water in a fifth microfuge tube. Add 2 μl of 6× sample buffer and mix.
3. Spin all five tubes for 10 seconds on high speed.

4. Load the standards and mini-prep DNA into sample wells in the agarose mini-gel. Electrophorese the gel in buffer at 100V until the fastest moving dye front has travelled approximately two-thirds of the distance down the gel (approximately 30 minutes).

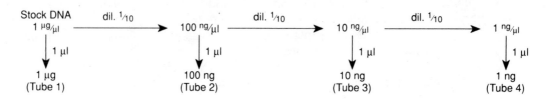

Stock DNA
1 $\mu g/\mu l$ — dil. $1/10$ → 100 $ng/\mu l$ — dil. $1/10$ → 10 $ng/\mu l$ — dil. $1/10$ → 1 $ng/\mu l$

↓ 1 μl ↓ 1 μl ↓ 1 μl ↓ 1 μl

1 μg 100 ng   10 ng   1 ng
(Tube 1)   (Tube 2)   (Tube 3)   (Tube 4)

5. Place the gel on a UV light box and compare the intensity of the mini-prep DNA band with the standards. (**Caution:** wear safety goggles and cover exposed skin when working around UV light.) Alternatively, a photograph of the gel can be taken if the equipment is available.
6. Estimate the amount of DNA in the band and determine the concentration of the mini-prep DNA sample. For example, if the intensity of the mini-prep DNA band most closely matches that of the 100 ng concentration standard, then the concentration would be 100 ng/2 μl, or 50 ng/μl (0.05 mg/ml).
7. Determine the total quantity (in μg) of mini-prep DNA obtained.

NOTES AND CALCULATIONS

Chapter 5

Introduction

Recombinant DNA molecules usually carry cloned genes that may encode proteins of interest to cell biologists. Conventionally, a cloned gene is often defined by its location relative to specific regions, called restriction sites, on a plasmid. Therefore, one of the first steps in characterizing a recombinant molecule is to create a restriction map. The information obtained from the restriction map, in conjunction with the data obtained from genetic experiments, provides researchers with a comprehensive picture of the recombinant molecule.

Plasmids are made of double stranded DNA, and therefore can be cleaved by enzymes called restriction endonucleases. Restriction enzymes are naturally occurring bacterial enzymes that recognize specific four to eight base pair sequences in DNA and make double stranded breaks somewhere within the sequence. Over 250 enzymes, each having a different recognition sequence, have been purified from a variety of bacterial strains, and can be purchased commercially from a number of different companies. Restriction reactions *in vitro* have strict reaction condition requirements, such as temperature, pH, and especially salt concentration, that differ among enzymes. Most restriction enzymes work over a very narrow concentration range of sodium chloride and pH and therefore come from the manufacturer with an accompanying 10× strength reaction buffer that provides the optimal salt concentration and pH for each enzyme.

The recognition sequences for restriction enzymes are usually palindromic, that is, they are the same when viewed backward or forward. Some restriction enzymes cleave at their recognition sequences in a symmetrical fashion, while others make staggered cuts in the DNA to yield single stranded overhangs, called cohesive or sticky ends. Sticky ends are able to base pair with compatible sticky ends from any DNA cut with the same enzyme. Therefore, a recombinant DNA molecule can be made by joining the compatible sticky ends generated when two entirely different types of DNA are cut with the same restriction enzyme. This technology has allowed scientists to clone a number of genes from eukaryotic sources, including human genes.

When a molecule of DNA, such as a plasmid, is cut with restriction endonucleases, a number of "restriction fragments" are generated. This process is similar to cutting a circular piece of string with scissors. If the string is cut once, a long linear fragment is obtained that represents the full length of the circle. Cutting the string more than once yields a number of shorter pieces whose sizes represent the distance between each of the sites cut with the scissors. When this is done with plasmid DNA, the sizes of the fragments represent the distance between restriction sites.

Restriction mapping analysis provides a means of locating restriction sites and their positions relative to one another on plasmid DNA. A restriction map can be constructed by cleaving the DNA with one or more restriction enzymes to generate a series of fragments. The fragments are then subjected to agarose gel electrophoresis and separated according to size. The length of each fragment generated by the restriction digestion is determined by comparing them to restriction fragments with known molecular sizes run on the same gel, such as bacteriophage λ DNA cut with the enzyme *Hin*d III. Once the sizes of the fragments are known, they can be pieced together into a comprehensive map.

Constructing restriction maps is an intuitive process. For example, consider the hypothetical plasmid pLAB. Restriction digestions of pLAB with the enzymes *Sal* I and *Bam*H I, and a double digestion with both enzymes generates the following fragments:

pLAB/*Sal* I	one fragment—9000 bp
pLAB/*Bam*H I	two fragments—6000 and 3000 bp
pLAB/*Sal* I, *Bam*H I	three fragments—4000, 3000, and 2000 bp

The enzyme *Sal* I cleaves the plasmid at a single site on the plasmid, to yield a single 9 kb band. This represents the full linear length of the plasmid, as shown below:

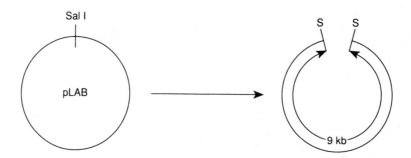

*Bam*H I cuts the plasmid twice to generate two fragments, one that is 6 kb, and one that is 3 kb long:

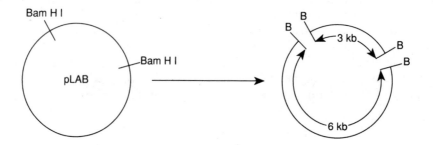

How are these restriction sites oriented in regard to one another? To answer this question, a double digestion with both *Sal* I and *Bam*H I is performed, to yield three fragments. Comparison of the fragments from the single digest to those from the double digest shows that the 3 kb *Bam*H I fragment from the single digest is retained in the double digest. This indicates that the *Sal* I restriction site lies somewhere within the 6 kb *Bam*H I fragment. *Sal* I cleaves the 6 kb fragment into two pieces of 4 kb and 2 kb:

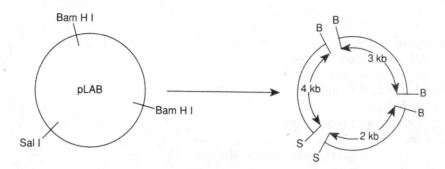

By convention, the unique restriction site is usually placed at the top of the map and called site 0. Other restriction sites are then placed on the map in a clockwise manner. In our example, then, the restriction map could be written:

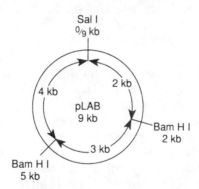

In this experiment, you will digest the pRPC245 mini-prep DNA obtained in the previous experiment with the enzymes *Eco*R I, *Pst* I, and *Sal* I. You will also digest bacteriophage λ DNA with *Hind* III for molecular size standards. You will then separate the restriction fragments by agarose gel electrophoresis. You will determine the sizes of the restriction fragments from pRPC245 by comparing them with the molecular size standards and construct a comprehensive map of pRPC245 from the restriction digestion data.

Materials

Stock λ DNA
Mini-prep DNA from previous experiment
Restriction enzymes and 10× reaction buffers:
   *Hin*d III
   *Eco*R I
   Pst I
   Sal I
Ice
Sterile distilled water
Agarose (electrophoresis grade)
50× TAE
Ethidium bromide (10 mg/ml)
6× loading buffer
Mini-gel electrophoresis apparatus
Power supply
Micropipettors (capable of pipetting less than 10 μl)
Sterile tips for micropipettor

Note to Instructors

Students should use the mini-prep DNA obtained from the previous experiment for the restriction digestion. Quantities of plasmid DNA should be prepared prior to this experiment in case student yields are low. This can be done by scaling up the mini-prep procedure described in this experiment or by using a commercial kit (see appendix II). Other plasmids and restriction enzymes can be substituted if they are available in your department.

Restriction enzymes and plasmid DNA must be kept on ice at all times. Students will be working with extremely small quantities of solutions, and the concentration of each component in the restriction reaction is critical. Therefore, students should practice using the micropipettors and microcapillary pipettes by pipetting with water until they feel confident enough to begin the digestion reaction.

Instead of having each student or student group run all of the reactions, the experiment may be set up so that each group prepares a single digestion (for example, pRPC245/*Eco*R I). When all of the digestions are complete, they can be loaded onto a single gel. Each student then gets a copy of the results to complete on his or her own.

This experiment can be split into two parts if necessary: (1) Set up and incubate the restriction digestions for 1–2 hours and (2) pour the agarose gel and electrophorese the samples. If this format is chosen, the reaction tubes should be heated in a 65° C water bath for 5 minutes following the restriction reaction to inactivate the enzyme. The digestions can then be stored at 4° C until electrophoresis is carried out.

Procedure

A. Restriction Digestion of Plasmid DNA

1. Prepare experimental protocols for the digestions shown below:

Tube	dH$_2$O	DNA	10× Buffer	Enzyme	Vol.
1	_____	5 μg λ	2 μl	10 U Hind III	20 μl
2	_____	1 μg pRPC245	2 μl	5 U EcoR I	20 μl
3	_____	1 μg pRPC245	2 μl	5 U Pst I	20 μl
4	_____	1 μg pRPC245	2 μl	5 U Sal I	20 μl
5	_____	1 μg pRPC245	2 μl	5 U EcoR I, Pst I	20 μl
6	_____	1 μg pRPC245	2 μl	5 U EcoR I, Sal I	20 μl
7	_____	1 μg pBR322	2 μl	5 U Pst I, Sal I	20 μl

The amount of sterile water added to the reaction is dependent upon the amount of enzyme and DNA added to make up the correct concentrations. For example, if the concentration of the λ DNA is 1 mg/ml (1 μg/μl) and the concentration of Hind III (according to the manufacturer) is 10 U/μl, then the protocol for the λ/Hind III digestion would be written as follows:

Tube	dH$_2$O	DNA	10× Buffer	Enzyme	Vol.
1	12 μl	5μl (5 μg)	2 μl	1 μl (10 U)	20 μl

2. Add the reactants to labelled microfuge tubes in the following order: water, 10× buffer, DNA, and enzyme. Thoroughly mix the contents by flicking the tubes with your fingers. Spin the tubes for 10 seconds in a microfuge at high speed.
3. Incubate the tubes in a 37° C water bath for at least 1 hour.
4. While the digestions are incubating, prepare a 0.7% agarose mini-gel, and 250 ml of running buffer (described in the previous experiment). When the gel has hardened, carefully remove the comb and place the gel into the electrophoresis apparatus. Add the running buffer to the buffer chambers.
5. After an appropriate incubation period, remove the restriction digestions from the water bath and place them on ice for 1–2 minutes.
6. Add 4 μl of 6× loading buffer to each tube and thoroughly mix.
7. Using a micropipettor set at 24 μl, carefully load the samples into wells in the agarose gel.
8. Electrophorese the gel at 100 V until the fastest moving dye front has travelled approximately two-thirds of the distance down the gel. Place the gel on a UV light box to visualize the DNA bands. A picture should also be taken at this time.

B. Determination of Fragment Size

9. Prepare a standard curve using the λ/Hind III digestion as molecular weight standards, as follows:
   a. Measure the distance d (in cm)' from the sample well to the top of each band in the size standard lane.

    b. Calculate 1/d for each of the measurements.

    c. The λ/*Hin*d III digestions generates fragments of known sizes (23130, 9416, 6682, 4361, 2322, 2027, 564, 125 bp). Plot size versus 1/d for each of the standards on linear graph paper.

    d. Draw a best fit line through the data points. It is important to note that the migration of macromolecules through an agarose matrix is a logarithmic process. Therefore, the largest fragments and the smallest ones will not migrate in a linear fashion. Omit them when drawing the best fit line.

10. Using the standard curve, determine the sizes of all of the fragments generated for each restriction digestion of pRPC245 by calculating 1/d for each band and reading the corresponding size from the standard curve.

11. Construct a restriction map of pRPC245 that fits the restriction data.

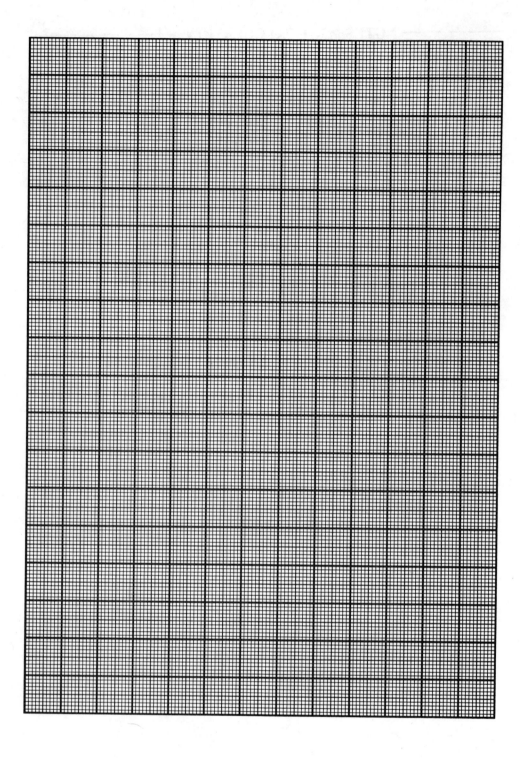

Chapter 6

Protein Fractionation: The Purification of IgG from Human Serum

Introduction

Cell and molecular biologists are often concerned with separating and analyzing macromolecules from cells; nucleic acids and proteins, for example. The nucleic acids, DNA and the RNAs, store and carry genetic information and direct protein synthesis by means of their linear nucleotide sequences. Proteins, the products of gene expression, carry out numerous essential cellular functions. Structural proteins determine the overall shape and appearance of cells. Other proteins are involved in molecular recognition. Enzymes, a major class of cellular proteins, serve as catalysts to speed chemical reactions. In order to learn more about the biological function associated with a particular molecule, it must first be separated from other cellular constituents and then purified for use in biochemical and biophysical assays.

Throughout the years, researchers have developed a number of different techniques for the purification of a specific protein from a protein mixture for further study. Some of these techniques are preparative (used to purify large quantities of a specific protein), while others are analytical (to identify or characterize proteins). Usually, the starting material for these separations is a crude cell or tissue homogenate. The separation of a single protein from such a mixture of thousands is a considerable task. It is often necessary to use a combination of techniques, which separate proteins on the basis of physical characteristics, such as size, charge, or affinity for a particular ligand.

Centrifugation is the simplest and crudest method that can be used to separate cellular components into different fractions. Subcellular fragments in a cell lysate can be separated by centrifuging at different speeds for fixed lengths of time since particles of different size travel toward the bottom of a tube at different rates in a centrifugal field. The soluble supernatant remaining after the highest speed centrifugation can provide the starting material for the purification of a soluble protein. A cleaner separation can be achieved by adding a density gradient to the centrifuge tube along with the sample to be fractionated. The particles in the sample will move until they reach a place in the tube that is equal to their own density.

Proteins can also be selectively precipitated out of solution by taking advantage of the solubility differences among different proteins. As the ionic strength of a protein mixture is raised (by the addition of a salt such as ammonium sulfate), different proteins will precipitate out of the solution when they reach their solubility limit. This process is called salting-out and is a popular first step in protein purification because it can be carried out on a large scale.

Column chromatography techniques have provided researchers with a means to fractionate proteins cleanly and in large quantities. In this technique, a heterogeneous protein mixture is passed through a column containing a solid matrix material. The rate at which different proteins pass through the column is dependent upon their interaction with various matrices that have been developed to discriminate between proteins on the basis of molecular size, net charge, or specific affinities. In affinity chromatography, for example, protein molecules are reversibly bound to the column matrix through ligands or antibodies to which they are specifically attracted. This powerful technique, in some instances, can bring about an essentially total purification of a desired protein in a single step.

A particularly useful method of column chromatography is ion-exchange chromatography, which separates proteins on the basis of net charge. The charge carried by any given protein is the summation of the charges of its individual amino acids. The charges of the amino acids are in turn determined by the nature of their side chains and the pH of the medium. At a given pH, called the isoelectric point, the number of positive charges will equal the number of negative charges, and the protein will be neutral. This is shown for the dipeptide, aspartate-lysine, in figure 6.1. At pHs above a protein's isoelectric point, the net charge will be negative. Conversely, the net charge on a protein will be positive if the pH of the solution is lowered below its isoelectric point. The charged proteins will then be attracted to particles carrying the opposite charge.

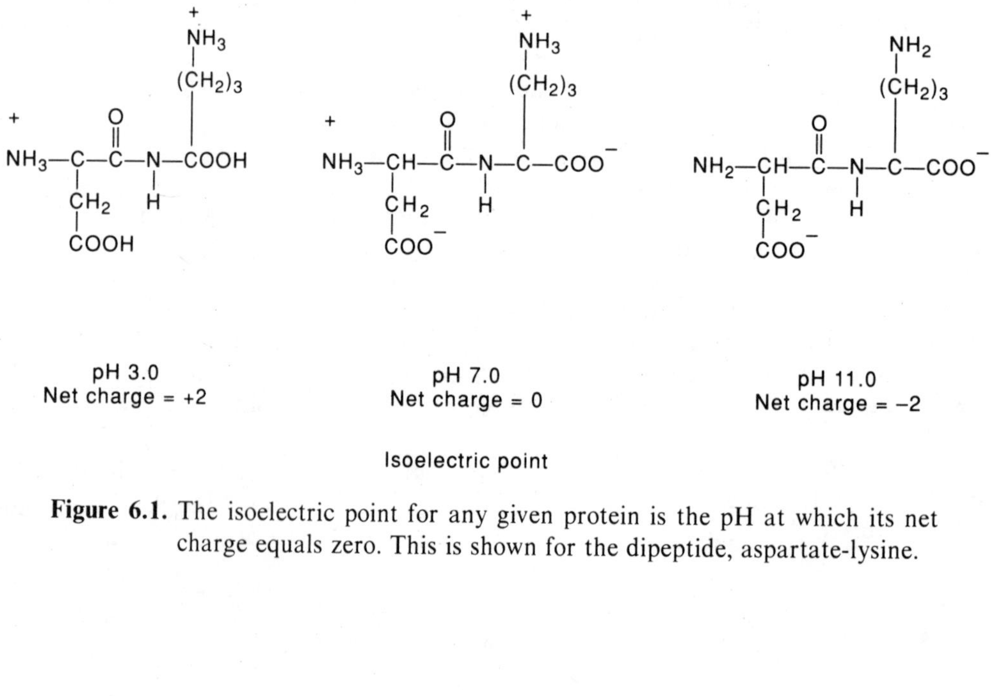

pH 3.0
Net charge = +2

pH 7.0
Net charge = 0

pH 11.0
Net charge = −2

Isoelectric point

Figure 6.1. The isoelectric point for any given protein is the pH at which its net charge equals zero. This is shown for the dipeptide, aspartate-lysine.

Ion exchange media consist of an insoluble matrix material to which charged functional groups are covalently bound. The functional groups can be positively charged (anion exchange media) or negatively charged (cation exchange media). Protein separations using ion-exchange chromatography are based on the principle that the charged particles that make up the column are able to reversibly adsorb oppositely charged protein molecules, while neutral or similarly charged proteins flow right through. The bound proteins can then be desorbed by increasing the ionic strength of the eluting buffer or by changing the pH, which changes the strength of the electrostatic interactions between the matrix and the sample molecules. This process is illustrated in figure 6.2. Ion-exchange chromatography leads to the sequential elution of bound molecules from the column in order of increasing affinity for the medium, until all of the molecules are eluted.

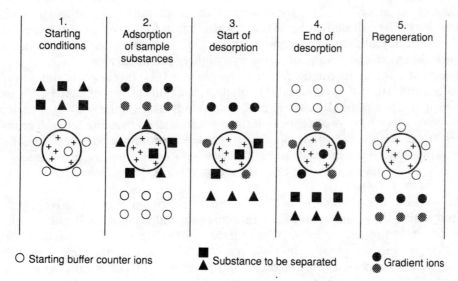

○ Starting buffer counter ions   ■ ▲ Substance to be separated   ● Gradient ions

Figure 6.2. Stage 1 shows the ion exchanger in equilibrium with its counter-ions. Sample substances are about to enter the ion exchange bed. In stage 2 the counter-ions have been exchanged for sample substances. After this adsorption a gradient is applied. Desorption of one sample species occurs at stage 3. This substance is exchanged for counter-ions in the eluting buffer and is therefore eluted from the ion exchanger. At stage 4 the remaining sample substance is exchanged for gradient ions and eluted, after which regeneration may be started. The gradient ions are exchanged for counter-ions in stage 5 and the ion exchanger is thus regenerated and ready for reuse. From Ion Exchange Chromatography Principles and Methods. Copyright © Pharmacia Inc., Piscataway, NJ. Reprinted by permission.

In this experiment, you will apply ion-exchange chromatography techniques to purify a single protein from a heterogeneous biological fluid, blood. The liquid fraction of blood, the plasma (or serum if clotting factors are removed), contains a multitude of different proteins that carry out numerous functions. The two proteins present in the greatest concentration are albumin, and the immunoglobulins (Ig's). The immunoglobulins, also called antibodies, are probably best known for their role in the immune response. Antibodies are highly specific biological macromolecules that recognize and bind to foreign substances (antigens). Although not actually lethal in themselves, they serve to recruit the cells of the immune system and other serum proteins, which then act in concert to destroy the offending particle. Antibodies can also combine with viruses or bacterial toxins and render them inactive.

There are five classes of immunoglobulins—IgG, IgM, IgA, IgD, and IgE. Only three, IgG, IgM, and IgA, have been well characterized, while little is known about the functions of the other two. The five classes differ structurally but possess two common features. First, they all share a basic structural subunit: four polypeptide chains hooked together with noncovalent interactions and disulfide bridges. Second, all immunoglobulins can recognize and bind to antigens in a highly specific fashion.

The most common immunoglobulin in the blood is IgG. It is also the simplest of the antibodies, consisting of two *heavy* (H) polypeptide chains of approximately 440 amino acids each and two *light* (L) chains, 220 amino acids long (figure 6.3). IgG molecules are bivalent—they have two identical antigen binding sites, enabling them to cross-link antigen molecules into a large insoluble network. This "immune complex" can then be processed by other immune system components. IgG antibodies are produced by activated B-lymphocytes as part of the *secondary* immune response, usually after a second exposure to an antigen.

At low ionic strength and neutral pH, most serum proteins are negatively charged. However, under these same conditions, IgG carries a neutral or slightly positive charge. These differences in net charge can be utilized to separate IgG from whole serum by ion-exchange chromatography. In this case, the anionic exchange medium DEAE-cellulose will be employed. This medium consists of a diethylaminoethyl group, which is a positively charged functional group (figure 6.4) covalently bonded to the insoluble support cellulose. The majority of serum proteins will adsorb to this medium at pH 7.2, while the positively charged IgG will not be bound.

Therefore, you can carry out this purification step on a large scale by *batching,* as opposed to passing it through a column containing the medium. You will stir the serum and ion-exchange medium together in a neutral buffer. You will then sediment the medium by centrifugation, along with the bound serum proteins. The supernatant contains purified IgG, which you will collect and assay for protein content. You will then store the purified IgG and later analyze it for purity and molecular weight by SDS-polyacrylamide gel electrophoresis (SDS-PAGE).

Materials

Dialysis tubing and clamps
Human serum (Sigma)
1 M phosphate buffer, pH 7.2
0.01 M phosphate buffer, pH 7.2

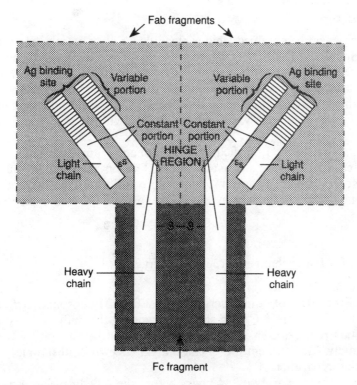

Figure 6.3. A schematic diagram of a molecule of IgG. From Bryant J. Neville, *An Introduction to Immunohematology,* 2d ed. Copyright © 1982 W. B. Saunders, Philadelphia, PA. Reprinted by permission.

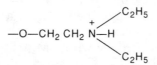

Figure 6.4. The positively charged functional group diethylaminoethyl is used in anion exchange.

Whatman DE52 ion-exchange medium
15 ml culture tubes
5 ml serological pipettes
Spectrophotometer
Pierce Protein Assay Reagent (Pierce)
2 mg/ml BSA concentration standard (Pierce)
Sterile distilled water
Micropipettors
Sterile micropipettor tips

Note to Instructors

Human serum can be obtained from a blood donor or by purchasing it from a company, such as Sigma, that sells cell culture reagents. If you use fresh blood, it should be collected in tubes without an anticoagulant, allowed to clot for at least 1 hour, and then centrifuged. The serum (straw-colored liquid on top of the clotted blood) is removed and can be stored in the refrigerator for up to 1 week. You should take precautions (wear gloves, autoclave contaminated waste) when working with human serum, regardless of the source.

Before beginning the IgG purification, the human serum must be dialyzed against phosphate buffer at pH 7.2 for at least 24 hours. The ion-exchange medium also needs to be washed several times in the same buffer to equilibrate it to the appropriate pH. This can be done by the students or by the instructor, as described in steps A.1 and A.2 below.

Procedure

A. Preparation for Ion-exchange Chromatography

1. Dialysis of human serum

Dialyze 30 ml of human serum against 0.01 M phosphate buffer pH 7.2 for at least 24 hours at 4° C with stirring, changing the buffer at least once. (Each separation requires 2 ml of serum; 30 ml should be enough for a whole class. Alternately, each student group can dialyze its own 2 ml sample).

2. Medium preparation

Begin the preparation 3–4 hours before the scheduled lab. Equilibration of 25 g of medium should provide enough for an entire class. This procedure can be scaled up or down if desired.

a. Stir 25 g of Whatman DE-52 medium into 500 ml of 1 M phosphate buffer, pH 7.2 (20 ml buffer for each dry gram of medium used). Allow the slurry to settle for 10 minutes, then remove and discard the supernatant.

b. Stir 500 ml more buffer into the medium. Allow it to settle, then remove and discard the supernatant, as in part a.

c. Continue to wash the medium with 1 M phosphate buffer pH 7.2 until the supernatant has a pH of 7.2. This may require several changes of buffer.

d. When the pH reaches 7.2, wash the medium 2 times with 0.01 M phosphate buffer pH 7.2. Leave the medium in excess buffer until just before the lab period begins, then remove most of the buffer. Mix the slurry completely just before use.

B. Separation of Serum Proteins by IEC

1. Transfer 10 ml of equilibrated ion-exchange medium to a 15 ml culture tube. Add 2 ml of dialyzed human serum, and mix completely by inverting the tube several times.

2. Incubate the tube at room temperature for 1 hour, with repeated mixing. During this incubation period, prepare a BSA standard curve, which you will use to determine the concentration of protein in the purified IgG solution (see Standard Curve for Protein Concentration Determination, step c).

3. Following the incubation period, centrifuge the tube at 1,000 g for 10 minutes. Remove the supernatant and place it into a clean 15 ml culture tube labelled "IgG". The pelleted ion exchange medium containing the bound serum proteins can be discarded.

4. Remove 0.5 ml of the purified IgG solution, and place in a small tube. Store the remainder of the sample at 4° C for use in chapter 7.

C. Standard Curve for Protein Concentration Determination

NOTE: The concentration of bovine serum albumin (BSA) standard protein solution is 2 mg/ml.

1. Turn on a spectrophotometer and set the wavelength at 595 nm. Allow the machine to warm up for at least 5 minutes. Use distilled water to set zero.

2. Label 6 13 × 100 glass tubes #1–6.

3. Dilute the BSA standard protein solution as shown below:

Tube	H_2O	BSA
1	50 µl	—
2	47.5 µl	2.5 µl
3	45 µl	5 µl
4	40 µl	10 µl
5	32.5 µl	17.5 µl
6	25 µl	25 µl

4. Add 2.5 ml of Pierce Protein Assay reagent to all of the tubes. Cover them with parafilm, and mix thoroughly by inversion.

5. Read and record the A_{595} for all of the tubes (note that tube #1 contains no protein. Tube #1 is therefore considered the "blank").

6. For tubes #2–6, subtract the A_{595} of the blank (tube #1) from the reading.

7. To create a standard curve, plot the net absorbance (A_{595} Sample—A_{595} Blank) of each sample against the amount of protein on linear graph paper, as shown below.

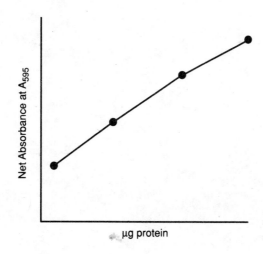

D. Determination of Protein Concentration in the Purified IgG Sample

   1. Label 5 13 $\times$ 100 glass tubes #1–5.
   2. Prepare dilutions of the purified IgG sample as shown below:

Tube	H_2O	IgG
1	50 μl	—
2	49 μl	1 μl
3	47.5 μl	2.5 μl
4	45 μl	5 μl
5	40 μl	10 μl

   3. Add 2.5 ml of Pierce Protein Assay Reagent to each tube, and thoroughly mix.
   4. Read and record the A_{595} of each of the tubes.
   5. Subtract the A_{595} of the blank (tube #1) from the readings for tubes #2–5.
   6. Using the BSA standard curve, determine the protein concentration of the purified IgG sample. For example, from the standard curve, an A_{595} reading corresponds to 10 μg of protein. If 5 μl of IgG sample were added to the protein assay reagent, the protein concentration would be 10 μg/5 μl, or 2 μg/μl (2 mg/ml).

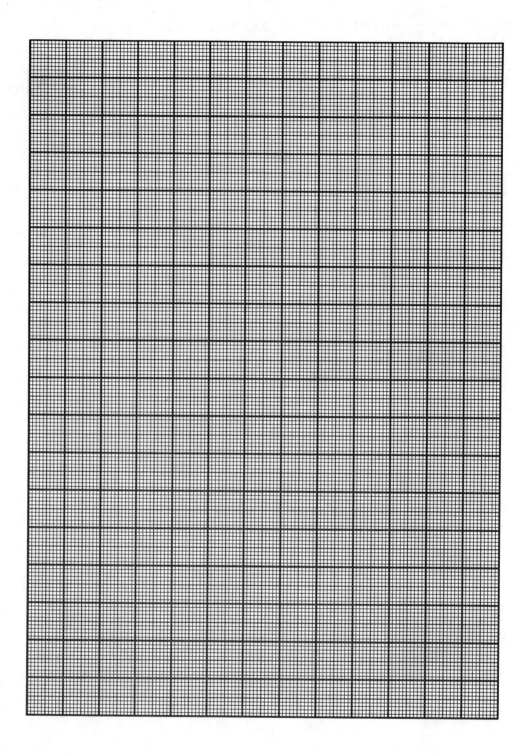

Chapter 7

Estimation of Protein Molecular Weight by SDS-PAGE

Introduction

The separation of a single protein from a complex mixture such as a cell extract is a crucial first step in studies on protein biochemistry and function. Gel electrophoresis is a very powerful tool used to fractionate proteins for analytical studies. Electrophoresis refers to the movement of charged molecules in an applied electric field. How fast a molecule moves is dependent upon the strength of the electric field, the net charge on the protein, and the viscosity of the medium through which it travels.

Electrophoretic separations of biological macromolecules are almost always performed in gels made of a porous insoluble material such as agarose or acrylamide. For separations involving proteins, acrylamide is the medium of choice because it is inert and, therefore, does not interact appreciably with the samples. Polyacrylamide gels are formed by reacting acrylamide and a cross-linking reagent, methylenebisacrylamide (bis), in the presence of an initiator such as N,N,N′,N′-tetramethyl-ethylenediamine (TEMED) and a catalyst for polymerization (amonium persulfate). The acrylamide monomers are cross-linked to form a polyacrylamide matrix. The average distance between the rod-like polymers is referred to as the pore size of the gel. Pore size can be varied by changing the concentrations of acrylamide and the cross-linking agent in the monomer solution (higher acrylamide concentrations result in the formation of smaller pores).

The polyacrylamide gel serves as a molecular sieve that separates proteins on the basis of their mass, with smaller proteins migrating more quickly and, thus, farther through the gel. Large molecules are impeded by the pores and migrate only a short distance (figure 7.1). In order to perform separations based solely on size, all of the molecules must have the same charge-to-mass ratio. This can be accomplished by electrophoresing the samples under denaturing conditions, using SDS-PAGE.

SDS-PAGE is the acronym for sodium dodecyl sulfate—polyacrylamide gel electrophoresis. SDS-PAGE is a simple and sensitive method used to fractionate proteins and provide an estimate of their denatured molecular weights. Protein samples are first boiled in a buffer containing sodium dodecyl sulfate, an anionic detergent, and a reducing agent such as β-mercaptoethanol or dithiothreitol, which breaks the disulfide bonds and disrupts the tertiary structure of the protein. The negatively charged SDS

50

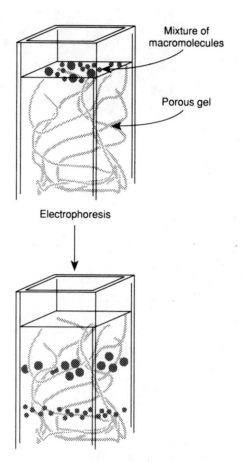

Mixture of
macromolecules

Porous gel

Electrophoresis

Figure 7.1. Sieving action of a porous polyacrylamide gel. From *Biochemistry,* third edition, by Lubert Stryer. Copyright © 1975, 1981, and 1988 by Lubert Stryer. Reprinted by permission of W. H. Freeman and Company.

molecules (figure 7.2) bind to the hydrophobic amino acid residues on the interior of the protein, disrupting the noncovalent interactions. The protein unwinds. Because of the mutual repulsion of the negatively charged sulfate groups on the SDS anions, the protein takes on a linear conformation (figure 7.3). The denatured proteins carry a large net negative charge approximately proportional to the molecular weight of the protein, a phenomenon that overwhelms the protein's native charge.

The denatured proteins are then loaded onto a polyacrylamide gel and electrophoresed. The negatively charged molecules migrate through the matrix toward the anode at a velocity that is roughly proportional to the logarithm of their molecular weight. How fast they travel is dependent on the pore size of the matrix and on the molecular weight of the protein.

Visualization of the proteins after electrophoresis is accomplished by first staining with Coomassie blue (silver stains can also be used), and then destaining with methanol and acetic acid. The protein bands are specifically stained, while the rest of the gel is destained, forming blue protein bands against a clear background as shown in figure 7.4.

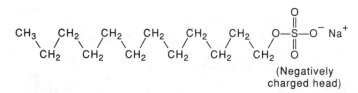

Figure 7.2. The anionic detergent sodium dodecyl sulfate (SDS).

Figure 7.3. Proteins denatured by heating in the presence of SDS and β-mercaptoethanol take on a linear conformation and are uniformly coated with negative charges.

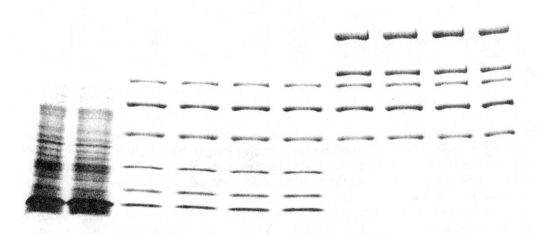

Figure 7.4. A Coomassie blue stained polyacrylamide gel. (Courtesy of Bio-Rad, Chemical Division)

52

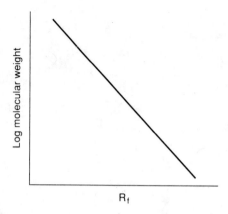

Figure 7.5. Protein standard curve of log molecular weight for a set of protein standards plotted against their relative mobility (R_f).

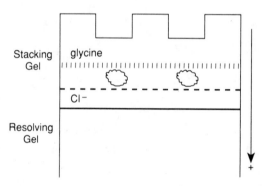

Figure 7.6. Proteins in a stacking gel are sandwiched between the glycine and chloride ion fronts.

Since the denatured proteins migrate through the gel on the basis of their size, it is possible to use SDS-PAGE to estimate molecular weight. Standard proteins of known size are electrophoresed on the same gel along with sample proteins. Molecular weight can then be determined by preparing a standard curve of relative mobility (R_f) versus the logarithm of molecular weight, as shown in figure 7.5.

In this experiment, you will analyze the IgG you purified from serum in Chapter 6 by SDS-PAGE for purity and in order to estimate the molecular weight of its polypeptide subunits. SDS-PAGE will be performed according to the Laemmli method, a two gel system that greatly improves resolution of proteins on polyacrylamide gels. The Laemmli system calls for a "discontinuous buffer system", which refers to the fact that two gels of different pH are employed. The upper, or stacking gel, is nonrestrictive to protein migration and serves to simply concentrate the protein samples between two ion fronts that are generated from the chloride and glycine ions in the electrophoresis running buffer. The proteins get stacked between the two ion fronts, as shown in figure 7.6.

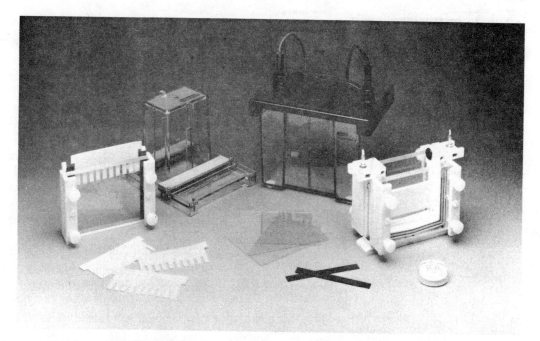

Figure 7.7. The Bio-Rad Mini-Protean II electrophoresus apparatus. (Courtesy of Bio-Rad, Chemical Division)

The mobility of these ions change when they reach the lower resolving gel at a higher pH. The glycine ions begin to migrate rapidly through the resolving gel, thus dissipating the protein "sandwich". The proteins are then separated on the resolving gel according to their molecular size.

Materials

Mini-Protean II electrophoresis equipment (Bio-Rad, Cat. #165–2940, see figure 7.7)
Mini-Protean II Ready Gels—12% single percentage gel (Bio-Rad, Cat. #161–0901)
Power Supply
Micropipettors
Micropipette tips
Microfuge tubes
IgG samples from the previous experiment
Sample Buffer
5X Running Buffer (dilute to 1X before use)
SDS-PAGE standards, 10,000–100,000 MW range (Bio-Rad, Cat. #161–0304)
Coomassie blue stain
Destain I (50% methanol, 5% acetic acid, freshly made)
Destain II (7% acetic acid, 5% methanol, freshly made)
Boiling water bath

Note to Instructors

The Mini-Protean II electrophoresis cell from Bio-Rad is sturdy and simple to use. Because of its small size, protein separations can be completed in less than 45 minutes. The premade Ready Gels that can also be purchased from Bio-Rad spares the preparer (either the students or the instructor) the time of pouring and polymerizing their own gels and also reduces the hazards associated with working with the neurotoxin acrylamide. There are many other mini-gel apparatuses that can be used in place of the Mini-Protean II with comparable results. A larger gel system, such as Hoeffer Scientific Instrument's Sturdier SE 400 vertical electrophoresis apparatus, can also be used. The recipe for a 12% polyacrylamide gel with a 4% stacking gel is provided in appendix III. Note that the polymerization of a large gel can take up to 4 hours, while a mini-gel can be polymerized in less than 1 hour.

Mini-gels will hold up to 10 samples; therefore, the entire class should be able to electrophorese their IgG samples on one or two gels.

In order to visualize the proteins, the gel must be stained with Coomassie blue, and then destained. This is usually a lengthy procedure but one which can easily be performed by student volunteers. To preserve the gel, it can be enclosed in a sealable plastic bag with a small amount of Destain II, or it can be dried down onto filter paper if the equipment is available. In either state, the gel can be photocopied to provide all of the students with their own copies of the results.

Assembly (according to manufacturer's instructions) and use of the equipment should be demonstrated. The students can then load their samples onto the prepared gel.

The SDS-PAGE molecular weight standards should be prepared and loaded onto the gel according to the manufacturer's instructions. In addition, 10 μl of a 1:100 dilution of human serum (mixed 1 μl to 4 μl with Sample Buffer) should be run on the gel to provide students with an appreciation of the plethora of proteins found in whole serum.

Procedure

1. Determine the volume of the IgG you purified in the previous experiment that would contain 50 μg of protein. For example, if you calculated that your protein yield was 5 mg/ml (5 μg/μl), you would need 10 μl of sample. Place the appropriate volume of IgG sample into a labelled microfuge tube.
2. Add 1/5 volume of Sample Buffer to the protein solution in the tube and thoroughly mix. For example, if the volume of your IgG sample is 10 μl, add 2.5 μl of Sample Buffer.
3. Place the tube in a boiling water bath and boil the samples for 2 minutes. Place the tube on ice.
4. Load the treated protein sample to the bottom of a well as demonstrated by your instructor, in the order shown below:

Lane 1	SDS-PAGE molecular weight standards
Lane 2	Diluted whole serum
Lanes 3–10	Student IgG samples

   Be sure to record which lane contains your sample.

5. Add Running Buffer (1X) to the buffer chambers of the electrophoresis apparatus, according to manufacturer's instructions. Connect the leads to a power supply, and electrophorese the samples until the bromophenol blue dye front has traveled to the very bottom of the gel (200 V for approximately 45 minutes using a mini-gel; 50 mA for 3 to 4 hours; or 10 mA overnight, using a large gel system).

6. After electrophoresis, remove the gel from between the glass plates, and submerge it in Coomassie blue stain for 1 hour.

7. Remove the gel from the stain solution and place in Destain I for 1 hour with gentle agitation.

8. Remove the gel from Destain I and place it into Destain II for 1 to 4 hours, until the blue protein bands stand out and the background is clear. Change the solution 2 or 3 times while the gel is being destained.

9. Once the gel is completely destained, you can enclose it in a sealable plastic bag with a small amount of Destain II, or dry it down onto Whatman 3MM filter paper.

10. To determine the molecular weights of the denatured polypeptides, plot a standard curve based on the linear relationship between the logarithm of molecular weight and the distance each polypeptide migrates from the top of the resolving gel:

    a. For each protein in the SDS-PAGE molecular weight standards, measure the distance in centimeters from the top of the resolving gel to the top of each band.

    b. Measure the distance from the top of the resolving gel to the top of the dye front band.

    c. Calculate the R_f (relative mobility) of each protein standard, using the following equation:

    $$R_f = \frac{\text{distance of protein migration}}{\text{distance of dye front migration}}$$

    d. On two cycle semi-log paper, plot R_f (on the linear scale) versus molecular weight (on the log scale). Draw a best fit line through the data points.

    e. To determine the molecular weight of any other polypeptide run on the same gel, calculate the R_f, and then read the molecular weight directly from the standard curve.

11. Determine the molecular weights of the polypeptide subunits of your purified IgG sample. What does this analysis tell you about the structure of the IgG molecule (remember that oligomeric proteins break into their component monomers when they are denatured and reduced)? From the gel, what is the estimated molecular weight of the complete IgG molecule? How does this compare to the actual molecular weight? Use the results of this experiment to determine if ion exchange chromatography can be used to efficiently purify IgG from whole serum.

——————————— NOTES AND CALCULATIONS ———————————

Chapter 8

Protein Biosynthesis: *in vitro* Translation

Introduction

Protein biosynthesis is an essential cellular function as inhibition of this process at any step results in cell death. The synthesis of proteins *in vivo* is a complex process—beginning with the transcription of an individual gene and ending with a polypeptide or protein. Scientists who first began studying this process shortly after the elucidation of the double helical nature of DNA quickly realized that the DNA could not serve directly as a template because in eucaryotic cells protein synthesis occurs in the cytoplasm. Genes, on the other hand, were found on chromosomes within the nucleus. Expression of the genetic material thus involved an intermediate—a messenger molecule that could carry the information to the cytoplasm.

Studies showed that the messenger was ribonucleic acid (RNA). RNA is chemically very similar to DNA, but contains the sugar ribose and a unique base, uracil, in place of thymine. The scientific evidence surrounding these molecules led Francis Crick to propose what he termed the "Central Dogma" of biology: that DNA is the template for the creation of new DNA molecules by DNA replication, and also for the synthesis of complementary strands of RNA (transcription). In turn, strands of messenger RNA (mRNA) are the templates for the synthesis of a colinear string of amino acids that makes up the primary structure of proteins (translation), as diagrammed in figure 8.1.

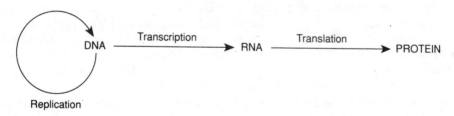

Figure 8.1. The Central Dogma of biology as proposed by Francis Crick.

60

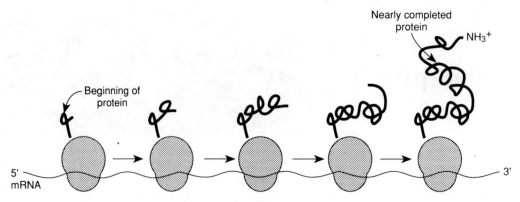

Figure 8.2. Translation of a strand of mRNA to yield a functional polypeptide. From *Biochemistry,* third edition, by Lubert Stryer. Copyright © 1975, 1981, and 1988 by Lubert Stryer. Reprinted by permission of W. H. Freeman and Company.

By the time the genetic code was fully deciphered by Marshall Nirenberg and H. Gobind Khorana in 1966, it had been shown that amino acids were assembled into proteins by the ribosomes, large cytoplasmic organelles made of protein and ribosomal RNA (rRNA). As the mRNA molecules emerge from the nucleus, ribosomes attach at a specific three base initiation "codon", AUG. Using energy derived from the hydrolysis of GTP molecules, the ribosomes proceed along the mRNA strand, and, with the help of adapter or transfer RNA molecules (tRNA), add amino acids to form a polypeptide chain (figure 8.2). The ribosomes function to hold the amino acids in proper alignment for the formation of peptide bonds. The chain always grows from the amino terminal amino acid toward the carboxyl terminus, corresponding to the codons in the mRNA strand. Newly synthesized proteins fold up into their native state as translation proceeds, so that the completed product is biologically active as soon as it is released from the ribosome.

One of the ways scientists chose to study protein synthesis and other cellular processes was to use extracts of disrupted cells. Cells contain the enzymes and other biochemical equipment necessary to make proteins, and it seemed logical that the process could be carried out *in vitro.* Once an active cell extract had been found (one that supported protein synthesis, if only to a limited degree), it was fractionated into its various components to determine the key players: the ribosomes, tRNA molecules, and initiation, elongation, and termination factors.

Fractionated cell (cell free) extracts played an important role in the elucidation of the genetic code. Synthetic mRNA molecules were added to cell extracts depleted of endogenous mRNA, and the synthesized proteins were analyzed for amino acid content. Initially, mRNA containing only one type of base was added. For example, the mRNA added might be a poly U-(UUUU. . . .) mRNA. The synthesized protein was found to consist of a string of phenylalanines. The codon UUU was thus assigned to the amino acid phenylalanine. In subsequent studies, repeating sequences were used (for example, GUUGUU . . .) to assign codons to all of the amino acids, thus fully deciphering the genetic code.

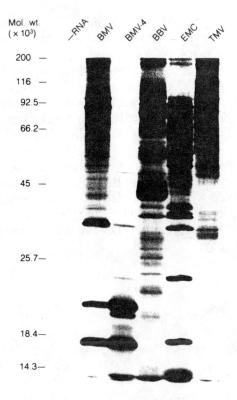

Mol. wt.
($\times 10^3$)

200 —
116 —
92.5—
66.2—
45 —
25.7—
18.4—
14.3—

Figure 8.3. Translation of a wide spectrum of mRNAs. The indicated mRNAs were translated in a 50μl reaction for 120 minutes and then analyzed by gel electrophoresis and autoradiography. RNAs translated were: Brome Mosaic Virus RNA (BMV); Brome Mosaic Virus RNA-4 (BMV-4); Black Beetle Virus RNA (BBV); Encephalomyocarditis Virus RNA (EMC); and Tobacco Mosaic Virus RNA (TMV). (Courtesy of Promega Corporation)

Since those historic studies over two decades ago, increasingly reliable cell free systems have been developed to synthesize proteins from exogenously added mRNAs. Purified extracts prepared from rabbit reticulocytes (red blood cell precursors) have been used successfully in many *in vitro* translation applications. These preparations, now available commercially, are treated to remove endogenous mRNA and contaminating ribonucleases that might degrade mRNAs added to the system. *In vitro* translation systems play an important role in identifying mRNA species isolated from various cells or tissues and in characterizing their protein products. These systems have also been successfully employed in studies on the control mechanisms for transcription and translation.

The protein products synthesized from exogenous mRNAs *in vitro* can be studied in a number of different ways. One particularly useful method is to add a radiolabelled amino acid to the system along with the other components. As protein synthesis proceeds, the labelled amino acid will be incorporated into the growing chain, forming a radio-active protein. The extract is then subjected to SDS-PAGE, which separates all of the proteins in the solution by mass. Following electrophoresis and staining, the polyacryl-amide gel is dried onto filter paper and placed into a film cassette along with a sheet of X-ray film (autoradiography). The radioactivity in the labelled band(s) exposes the film, and when the film is developed, dark bands are seen which correspond to the location of the labelled proteins (figure 8.3). Thus, only newly synthesized proteins will be detected on the autoradiogram. The other proteins from the cell extract are seen on the stained gel only.

In this experiment, you will use a commercially prepared rabbit reticulocyte *in vitro* translation system to translate RNA from Brome Mosaic Virus (BMV) in the presence of ^{35}S-methionine. BMV is a single stranded RNA virus whose genome encodes four proteins (a fifth translational product related to the viral coat protein may be seen as well). The proteins will be separated by SDS-PAGE, and the synthesized products visualized by autoradiography.

Material

Mini-Protean II
Mini-Protean II Ready Gels
Power supply
Gel drying equipment
Sample Buffer
Coomassie blue stain
Destain I and II
5X Running Buffer (dilute to 1X before use)
Rabbit Reticulocyte Lysate Nuclease Treated (Promega—L4161)
^{35}S-methionine (New England Nuclear)
Brome Mosaic Virus RNA (Promega—D1541)
Sterile distilled water
Microfuge tubes
Micropipettor and sterile tips
30° C water bath
67° C water bath
Boiling water bath
X-ray film (Kodak X-OMAT AR)
Film cassette (Sigma)
Developing reagents (developer and fixer—Kodak)
Rubber gloves
Plexiglass shielding (optional)

Note to Instructors

This experiment requires the use of the radioisotope ^{35}S. This isotope is considered a low energy emitter of β-particles, but should nonetheless be treated with caution and respect. Make sure that the students fully understand the minimal hazards associated with its use. Students should wear gloves and a lab coat when working with the isotope. Plexiglass shielding is also recommended.

Because ^{35}S is such a weak emitter, the polyacrylamide gel must be dried down onto either Whatman 3MM paper or a mylar sheet before exposure to the X-ray film. Gel drying equipment is likely to be available in a well equipped biology department. Alternately, gel drying frames, for drying gels onto mylar sheets, can be purchased from a number of suppliers (Hoefer, for example) at reasonable cost. They can also be used to dry down the polyacrylamide gel in chapter 7. Once the gel has been dried and used for autoradiography, it can be photocopied for distribution to the students, reducing their exposure to the radioactive gel.

After exposure to the dried gel, the X-ray film is developed using standard photographic methods (5 minutes in developer, 1 minute rinse in water, 5 minutes in fixer) performed in the dark. Darkroom facilities are an integral part of most biology departments; however, it is quite simple to purchase the solutions (from Kodak) and develop the film in a makeshift darkroom if necessary. It is advised that the instructor dries down the gel and performs the autoradiography, which will take approximately two days to complete. The autoradiogram can be photocopied for distribution.

Since each student group prepares one sample, one (or two) gel(s) should be sufficient for an entire class. After the samples are boiled in preparation for SDS-PAGE, they should be placed on ice until the entire class is ready to load the gel.

Just before the lab begins, the *in vitro* translation system reagents should be removed from the freezer and thawed on ice. Two control tubes containing (1) lysate, water, amino acid mix, and isotope, and (2) lysate only, should be prepared for SDS-PAGE (one per class) by a student volunteer or by the instructor.

Procedure

1. Set up SDS-PAGE mini-gel equipment according to manufacturer's instructions (see chapter 7).
2. Place 2 μl (or the volume required to make 1 μg) of BMV-RNA into a microfuge tube. Heat the tube in a 67° C water bath for 10 minutes, and then immediately cool on ice. This treatment denatures regions of secondary structure in the RNA and increases translation efficiency.
3. Add the following reagents to the heated BMV-RNA:

   35 μl rabbit reticulocyte lysate
   7 μl sterile distilled water
   1 μl 1 mM amino acid mixture (without methionine)
   5 μl ^{35}S-methionine (at 10 mCi/ml, or 50 μCi)

   ---

   50 μl—total reaction volume (final ^{35}S-methionine concentration should be 1 mCi/ml)

   ** **Note:** Gloves and a lab coat must be worn when working with the isotope. Discard all radioactive waste in a labelled container for proper disposal.

4. Carefully cap the tube and mix the contents by gently flicking it with your finger. Spin the tube briefly in a microfuge.
5. Incubate the tube in a 30° C waterbath for 1 hour.

6. Using a micropipettor, remove 10 μl from the reaction tube and place it into a new tube (discard the used micropipette tip and the reaction tube in the radio-active waste). Add 2.5 μl of SDS-PAGE Sample Buffer and gently mix. Boil the tube for 2 minutes, then immediately place it on ice.

7. Spin the sample for 30 seconds on high speed in a microfuge. Using a micro-pipettor, carefully load the polyacrylamide gel as follows:

Lane 1	SDS-PAGE molecular weight standards
Lane 2	Control tube 1 (no RNA added)
Lane 3	Control tube 2 (lysate only)
Lane 4–10	Student samples

If all of the samples do not fit on a single gel, run a second one at the same time. Each gel must contain the molecular weight standards and control tubes.

Discard the microfuge tube and the used pipette tip in the radioactive waste.

8. Add Running Buffer (1X) to the buffer chambers and connect the leads to the power supply. Electrophorese at 200V for 45 minutes (for a mini-gel—see chapter 7 for other gels) until the bromophenol blue dye front has migrated to the bottom of the gel.

9. Remove the gel from the electrophoresis apparatus and submerge it in Coomassie blue stain for 1 hour with gentle agitation.

10. Destain the gel in Destains I and II as described in chapter 7, until the background is clear and the protein bands prominent.

11. Dry the gel onto filter paper or a sheet of mylar. In the dark, place the dried gel into a film cassette along with a piece of X-ray film. Close and completely seal the cassette. Expose the film at room temperature for 12 to 24 hours.

12. Develop the X-ray film according to standard photographic procedures.

13. Compare the autoradiogram with the dried gel (**note:** the dried gel is still radioactive). Determine which proteins are the result of *in vitro* translation of the BMV-RNA.

14. Prepare a standard curve from the SDS-PAGE molecular weight standards on the dried gel, as described in chapter 7.

15. Calculate the R_f for each translational product and determine the molecular weights by reading them from the standard curve.

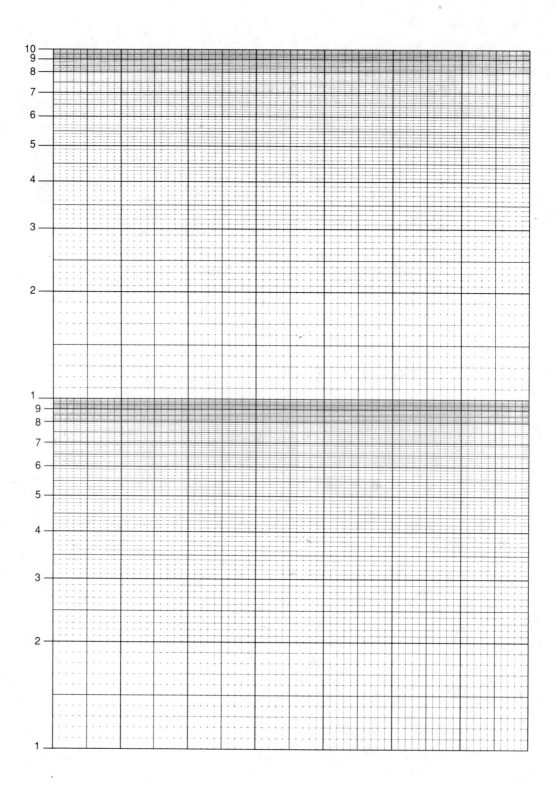

Chapter 9

Isolation of Mitochondria from Cauliflower Cells

Introduction

Aerobic respiration is the process used by oxygen-requiring organisms to generate energy required for essential cellular reactions. During respiration, high energy food sources like glucose are metabolized to generate chemical energy in the form of ATP, as summarized in the following equation:

$$C_6H_{12}O_6 + 6O_2 \longrightarrow 6CO_2 + 6H_2O + 38\ ATP$$
$$(energy)$$

The biochemical machinery required for cellular respiration is found in the mitochondria, small organelles scattered throughout the cytoplasm of eucaryotic cells. When viewed by light microscopy, mitochondria appear as simple rod-shaped bodies (figure 9.1). However, the electron microscope reveals the true complexity of the tiny structures (figure 9.2).

Each mitochondrion is bounded by two very specialized membranes, the *inner* and *outer* mitochondrial membranes, separated by the intermembrane space. The outer membrane surrounds the entire organelle and acts as its primary permeability barrier. The inner membrane is folded into numerous fingerlike projections called *cristae* that project into the *matrix,* a complex solution containing hundreds of enzymes and other proteins. The convoluted nature of the inner membrane serves to increase its surface area. This is an important feature because the proteins and enzymes responsible for cellular respiration are found here. A schematic representation of a mitochondrion is shown in figure 9.3.

Energy-producing organelles such as the mitochondria are unusual in that they have their own genetic systems that are entirely separate from the cells' genetic material. The mitochondria are, however, still dependent upon the cells' nuclear DNA to encode essential proteins required for mitochondrial replication, such as the polymerases and other replicative enzymes. Division of the organelle to yield more mitochondria requires input from the cells' genetic system but is for the most part not tied to the cells' replicative machinery. Mitochondria seem to divide randomly, out of phase with both the cell cycle and each other. While proteins encoded by nuclear DNA have been found in the mitochondria, mitochondrial proteins are not exported from the organelle.

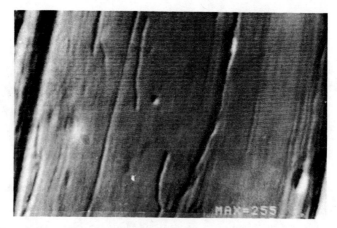

Figure 9.1. Mitochondria in a crayfish motor neuron, viewed by differential inter-ference contrast microscopy (DIC). (Photo courtesy of Christopher Case, State University of New York at Albany.)

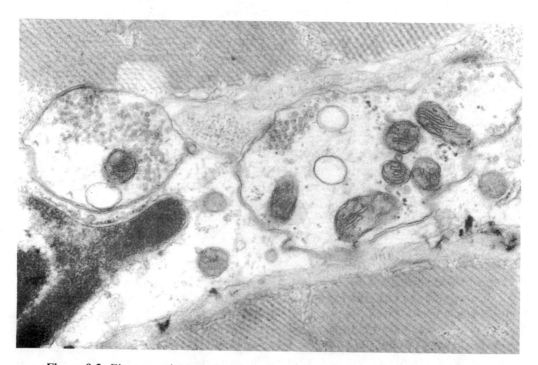

Figure 9.2. Electron micrograph showing mitocondria in crayfish axon terminals. (Courtesy of Bernard Ng, State University of New York at Albany.)

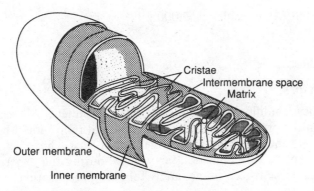

Figure 9.3. Schematic representation of a mitochondrion. From Stephen L. Wolfe, *Biology of the Cell.* Copyright © 1972 Wadsworth Publishing Company, Belmont, CA.

The nature of the mitochondrial genome and protein synthesis machinery has led many researchers to postulate that mitochondria may have arisen as the result of the ingestion of a bacterium by a primitive cell millions of years ago. In a theory known as the *endosymbiont hypothesis,* it is postulated that the two may have entered into a symbiotic relationship and eventually became dependent upon one another—the cell sustained the bacterium, while the bacterium provided energy for the cell. Gradually, the two evolved into the present day eucaryotic cell, with the mitochondrion retaining some of its own DNA. Because mitochondrial DNA is inherited in a non-Mendelian fashion in eucaryotic cells (mitochondria are inherited from the maternal parent, who supplies most of the cytoplasm to the fertilized egg), it has been used to look at evolutionary trees, including research on the origins of man.

In this experiment, you will isolate mitochondria from cauliflower cells. Cauliflower is used because, as a plant, it is nonhazardous and easy to work with. You will grind cauliflower rosettes and fractionate the homogenate by differential centrifugation. The resulting mitochondrial prep will then be frozen and stored for use in the following experiment.

Materials

Cauliflower (fresh—one head per 10 students)
Isolation Buffer (0.3 M D-mannitol, 0.02 M phosphate buffer, pH 7.2)
Assay Buffer (0.3 M D-mannitol, 0.02 M phosphate buffer pH 7.2, 0.01 M KCl, 0.005 M MgCl$_2$)
Purified sea sand
Cheesecloth
50 ml centrifuge tubes
15 ml sterile culture tubes
Razor blade
70% ethanol
Mortar and pestle

Centrifuge (refrigerated, capable of 12,000 g)
Glass slides
Coverslips
Phase-contrast microscope

Notes to Instructors

Use the freshest cauliflower available for this experiment. All reagents, tubes and containers used in the preparation of the mitochondrial prep should be kept cold at all times. Place mortars and pestles in the refrigerator at least 1 hour before the scheduled lab period, and keep the tubes and solutions on ice during the experiment. The mitochondrial prep should be frozen immediately following preparation.

Procedure

1. Remove several rosettes from a head of cauliflower. Using an ethanol-cleaned razor blade, slice off the outer 3–5 mm of each rosette, until approximately 25 g of tissue have been accumulated.
2. Place the sliced cauliflower into a chilled mortar. Add 6 g of cold sea sand and 20 ml of cold isolation buffer. Grind the tissue vigorously for about 2 minutes, until the mixture becomes a smooth paste.
3. Add an additional 20 ml of cold isolation buffer to the cell paste and grind for 2 minutes more.
4. Pour the slurry through four layers of cheesecloth. Collect the filtrate in a cold 50 ml centrifuge tube. Wash the mortar with 5 ml of cold isolation buffer and pour the wash through the cheesecloth as well. Squeeze out all of the juice into the tube.
5. Centrifuge the tube for 10 minutes at 600 $\times$ g in a refrigerated centrifuge set at 4° C.
6. Using a 10 ml serologic pipette, remove the postnuclear supernatant into a clean 50 ml centrifuge tube. Discard the pellet, which contains the nuclei.
7. Centrifuge the supernatant at 12,000 $\times$ g for 30 minutes at 4° C.
8. Remove the postmitochondrial supernatant from the mitochondrial pellet. Add 5 ml of the supernatant to a 15 ml culture tube labelled "postmitochondrial supernatant" to be frozen. The rest of the supernatant can be discarded.
9. Add 5 ml of cold assay buffer to the mitochondrial pellet remaining in the tube. Thoroughly resuspend the pellet in the buffer (this may require the use of a rubber spatula to scrape the pellet off the wall of the tube, and a pasteur pipette to completely disperse the clumps).
10. Transfer the suspension to a cold 15 ml culture tube labelled "mitochondria". Remove one drop and place it on a clean glass slide, and cover the drop with a coverslip.
11. Freeze the remaining mitochondria and the postmitochondrial supernatant at −20° C for use in the following experiment. The mitochondria can be stored frozen for up to 2 weeks.
12. Examine the mitochondria prep on the slide using a phase-contrast microscope at 45X and 100X. Note the concentration and appearance of the mitochondria.

Chapter 10

Electron Transport Chain— Assay for Enzyme Activity

Introduction

When a eucaryotic cell respires aerobically, it uses sugars and fatty acids as food and breaks them down to carbon dioxide and water. The chemical energy released following the oxidation of these substances by molecular oxygen is then captured and temporarily stored in the form of high energy phosphate bonds in the nucleotide adenosine triphosphate (ATP). This process occurs in the mitochondria.

ATP, shown in figure 10.1, has two phosphate bonds which when broken, release the stored energy that can then be used to power other vital cellular processes, such as macromolecular biosynthesis, ion transport, or motility.

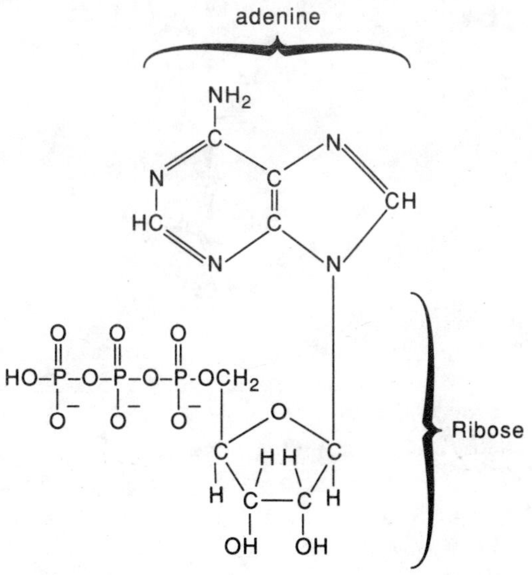

Figure 10.1. Energy is stored in the phosphate bonds of adenosine triphosphate (ATP).

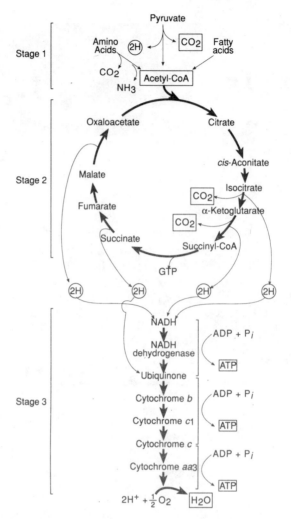

Figure 10.2. Stages in cell respiration. *Stage 1:* Mobilization of acetyl-CoA from glucose, fatty acids, and some amino acids. *Stage 2:* The citric acid cycle. *Stage 3:* Electron transport and oxidative phosphorylation. Each pair of *H* atoms entering the electron-transport chain as *NADH* yields *3 ATPs.* From A. Lehninger, *Principles of Biochemistry.* Copyright © 1982 Worth Publishers, Inc., New York, NY. Reprinted by permission.

In cells, ATP is synthesized following a series of phosphorylation steps—chemical reactions in which phosphate groups are sequentially added to a precursor molecule to create first ADP and then ATP. These phosphorylation steps require a large input of energy and are therefore coupled to an energy-generating system called the electron transport chain found in the inner membranes of the mitochondria.

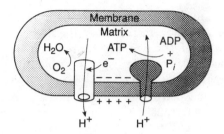

Figure 10.3. Oxidation and phosphorylation are coupled by the proton-motive force. From *Biochemistry,* third edition, by Lubert Stryer. Copyright © 1975, 1981, and 1988 by Lubert Stryer. Reprinted by permission of W. H. Freeman and Company.

One of the key features associated with cellular respiration as an energy-generating system is the gradual, controlled release of energy, rather than an explosive event where energy is wasted. This feat is accomplished via the transfer of electrons from one molecule to another in the electron transport chain. This chain consists of a series of electron-donating and electron-accepting substances, which pass along electrons until they reach the final electron acceptor, molecular oxygen (figure 10.2). As the electrons flow through the chain, small amounts of energy are released. This energy is used to pump protons (H^+ ions) from the inside to the outside of the inner mitochondrial membrane. As H^+ ions accumulate, a substantial pH gradient forms across the membrane. At the same time, a voltage gradient, or membrane potential, is produced. The combined effect, called the *proton-motive force* (figure 10.3), drives the H^+ ions back to the inside of the membrane, through special channels that are part of a large transmembrane protein complex, ATP synthetase. As the protons flow through the enzyme complex, the energy of the proton motive force is used to phosphorylate ADP. This energy is then stored in the phosphate bonds of the newly created ATP molecules. Because the phosphorylation of ADP is coupled to the oxidation of energy rich substrates like glucose, this process is called oxidative phosphorylation, and generates 28 of the 38 ATP molecules produced from the oxidation of 1 mole of glucose.

In this experiment, you will assay the activity of the enzyme succinate dehydrogenase, one of the enzymes involved in cell respiration. Unlike the other enzymes of the TCA cycle, succinate dehydrogenase and its coenzyme, flavin adenine dinucleotide (FAD), are bound to the inner mitochondrial membrane. Succinate dehydrogenase catalyzes the oxidation of succinate to fumarate in a reaction that generates two protons and two electrons, as summarized in the following reaction:

$$\text{Succinate} + \text{FAD} \xrightarrow[\text{dehydrogenase}]{\text{succinate}} \text{Fumarate} + \text{FADH}_2$$

The reduced form of the coenzyme ($FADH_2$) carries the protons and electrons to the electron transport chain, where ubiquinone accepts them and subsequently passes them down the chain to oxygen.

The activity of succinate dehydrogenase can be assayed by replacing ubiquinone with an artificial electron acceptor, in this case, the dye 2,6-dichlorophenolindophenol (DCIP). The oxidized form of the dye will be reduced by accepting electrons from $FADH_2$, with a corresponding change in color from blue to colorless. This change can be measured spectrophotometrically. To ensure that the electrons are passed to the dye and are not funneled into the electron transport chain, the poison sodium azide is added to the reaction to block the final transfer of electrons from cytochrome a_3 to oxygen. This effectively bottlenecks the electron transport chain, preventing $FADH_2$ from passing along its electrons to ubiquinone.

In this experiment, you will perform an enzyme assay on the mitochondrial prep and the post mitochondrial supernatant prepared in the previous experiment. You will determine the activity of the enzyme by relating the oxidation of succinate by succinate dehydrogenase to a decrease in absorbance resulting from the reduction of DCIP.

Materials

Spectrophotometer
13 × 100 glass tubes
1 and 5 ml serologic pipettes
Pasteur pipettes
Assay buffer (0.3 M D-mannitol, 0.02 M phosphate buffer pH 7.2, 0.01 M KCl, 0.005 M $MgCl_2$)
0.04 M sodium azide
50 μM DCIP
0.2 M succinate
Mitochondrial prep and postmitochondrial supernatant from the previous experiment, thawed on ice.

Note to Instructors

Sodium azide is a poison and should not be pipetted by mouth. The mitochondrial prep should be thawed on ice and kept in the ice bath at all times. Just before use in the enzyme assay, it should be thoroughly mixed by gently pipetting up and down several times with a chilled pasteur pipette.

The spectrophotometers should be allowed to warm up for at least 5 minutes before any readings are taken. Correct use of the instrument should be demonstrated to the students before they begin.

Procedure

A. DCIP Standard Curve
1. Label 6 13 × 100 tubes # 1–6.
2. Prepare dilutions of the 50 μM DCIP solution, as shown below:

Tube	Assay Medium	DCIP	Concentration
1	5 ml	—	0 μM
2	4 ml	1 ml	10 μM
3	3 ml	2 ml	20 μM
4	2 ml	3 ml	30 μM
5	1 ml	4 ml	40 μM
6	—	5 ml	50 μM

3. Cover each tube with parafilm and invert several times to mix.
4. Read and record the A_{600} of each tube, using tube #1 as a blank.
5. Draw a DCIP standard curve by plotting the A_{600} of each tube against the DCIP concentration on linear graph paper. Draw a best fit line through the data points, as shown below:

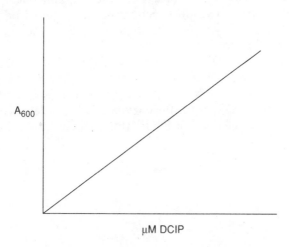

A_{600}

μM DCIP

B. Succinate Dehydrogenase Assay
   1. Label 6 13 × 100 tubes #1–6.
   2. To each tube, add assay buffer, sodium azide, succinate, and DCIP, as shown in the chart below. Cover each tube with parafilm and invert several times to mix thoroughly.

Tube	Buffer	Azide	Succinate	DCIP	MITO	SUP
1	3.5 ml	0.5 ml	0.5 ml	—	0.5 ml	—
2	3.0 ml	0.5 ml	0.5 ml	0.5 ml	0.5 ml	—
3	3.5 ml	0.5 ml	0.5 ml	—	—	0.5 ml
4	3.0 ml	0.5 ml	0.5 ml	0.5 ml	—	0.5 ml
5	4.0 ml	0.5 ml	0.5 ml	—	—	—
6	3.5 ml	0.5 ml	0.5 ml	0.5 ml	—	—

3. Using a pasteur pipette, gently mix the cold mitochondrial suspension and post-mitochondrial supernatant obtained in last week's experiment. Add 0.5 ml of mitochondrial suspension to tubes #1 and #2, and note the time (this is time 0). Immediately following the addition of the mitochondria, mix the tubes thoroughly and read the A_{600} of tube #2, using tube #1 as a blank. Place the tubes in a test tube rack at room temperature.
4. Add 0.5 ml of postmitochondrial supernatant to tubes #3 and #4, again noting the time. Immediately mix the tubes and read the A_{600} of tube #4, using tube #3 as a blank. Stand the tubes in the test tube rack with tubes #1 and #2.
5. Mix tubes #5 and #6 thoroughly, and read the A_{600} of tube #6, using tube #5 as a blank, and place these in the rack as well.

1 – 3.02 .07

2 – 0

3 0

4

6. At 5 minute time intervals, read the A_{600} for tubes #2 and #4. Remember to use the correct blanks to adjust the spectrophotometer. Take absorbance readings for 30 minutes. (Tube #6 is a control tube and only needs to be read at time 0 and time 30. The absorbance should not change since DCIP will not be reduced.)

7. Prepare a plot of A_{600} against time for the mitochondrial solution and the post-mitochondrial supernatant solution on linear graph paper (plot both on one graph). Compare the two lines. Since succinate dehydrogenase is membrane bound, most of the enzyme activity should be seen in the mitochondrial prep (as evidenced by the decrease in absorbance).

8. To further quantitate enzyme activity, use the DCIP standard curve to determine the concentration of DCIP present at each time point for the mitochondrial suspension, postmitochondrial supernatant, and for the control tube (#6). This should illustrate the activity of the enzyme succinate dehydrogenase, by relating it to the reduction of the dye DCIP (for each succinate oxidized, a DCIP is reduced).

Chapter 11

Growing Human Cells in Culture

Introduction

One of the biggest breakthroughs in the field of cell biology came in the early 20th century with the discovery that plant and animal cells, could survive—and even replicate—outside of the living organism. In 1907, neurobiologist R. G. Harrison in his quest to prove that nerve fibers were actually outgrowths of single cells chopped up spinal cord tissue and added to it clotted plasma in a humidified growth chamber. The nerve cells from this crude explant not only grew under these conditions but were also seen to extend long axons into the clot. This discovery clearly illustrated that cells could be grown in an artificial environment. Glass or plastic culture vessels eventually replaced plasma clots, and specialized growth media—containing balanced mixtures of amino acids, salts, vitamins and minerals—were developed to support the growth of cells *in vitro*.

With the advent of the biotechnology era of the 1970s, cell culture has taken on a whole new meaning. Genetically engineered mammalian cells grown in culture produce and secrete therapeutically important proteins such as interferon, growth hormones, and insulin. Hybridoma technology, the science of fusing immune mouse cells with immortal cancer cells, has made it possible to produce monoclonal antibodies from cultured cells. There are many advantages to growing cells in culture. Cell populations can be studied removed from the influences of other cells. Since an artificial environment is created, it is possible to study the complex behavior of cells in a strictly regulated environment that is completely under the control of the investigator. It should be noted, however, that cell culture is a two-sided coin—cells grown *in vitro* may behave differently than the same cells in their natural environment.

The principles behind cell culture are quite basic. Cells to be studied (for example, fibroblasts) are removed from representative tissues by dissociation with proteolytic enzymes. The cells are placed into a culture vessel to provide a growth surface, and supplied with the appropriate nutrients (in the culture medium), temperature, gaseous environment, and pH. Once the various growth requirements are satisfied, the cells begin to reproduce. As the population increases, the cells begin to release chemical signals into the medium to arrest reproduction. This is referred to as *contact inhibition,* and serves to prevent the culture from becoming overcrowded. A continuous, or *confluent,* monolayer of cells results that covers the entire growth surface of the culture vessel (figure 11.1).

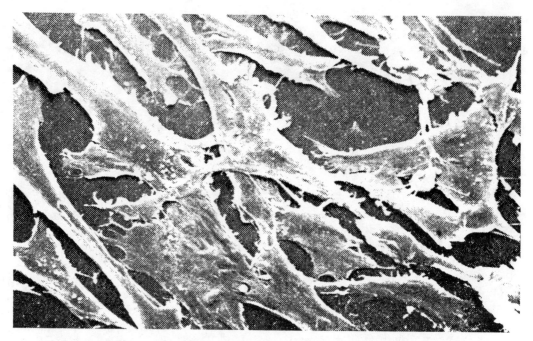

Figure 11.1. Electron micrograph of fibroblasts from a rat growing in culture. © Dr. G. Stephen Martin

Cell cultures that are derived directly from a particular tissue are called *primary cultures*. Primary cultures can then be subcultured to form *secondary cultures,* which can be maintained for long periods of time by repeated subculturing. These cultures cannot be maintained indefinitely, however, because cells in culture have a limited life span (as do cells in their natural state). Cultured cells undergo many of the same processes as they do *in vivo*. Fibroblasts, for example, continue to secrete collagen and fibronectin; nerve cells extend axons and synapse with other nerve cells, and so on.

Occasionally a mutation will occur within a population of cultured cells that results in virtual immortality for the culture. These cells are normal yet can be subcultured an infinite number of times. These types of cells can be propagated into a *cell line* and used in research studies over very long periods of time. Cancerous, or *transformed* cells, on the other hand, are not only immortal in culture but *in vivo* as well. Normal cells can become transformed as a result of exposure to chemicals or oncogenic viruses or by other less well understood means. In addition to morphological changes, transformed cells exhibit aberrant behavior—they can grow in suspension (without a solid surface) and do not exhibit contact inhibition. Cell division is also abnormal in transformed cells, and so in culture they appear to continuously divide, forming large clumps of cells that can be easily detected microscopically. These types of cells have been shown to cause cancerous tumors when injected into laboratory animals.

Cell culture is a subjective science (or is it an art?) and its practice varies among different laboratories (figure 11.2). In this experiment, you will be shown one method to grow human fibroblasts in culture over a three-week period, using standard cell culture techniques. You will evaluate the cultures during the second week. If they are viable, you will use them to test for cell adhesion in the next experiment.

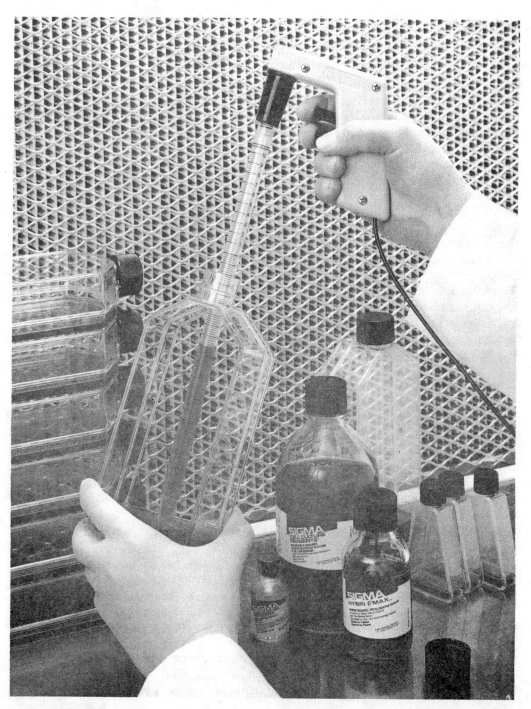

Figure 11.2. The art of cell culture. Sigma Chemical Company

Materials

The amount of culture medium, saline, and trypsin/EDTA solution required to supply a pair of students for the entire subculturing period is given.

One 25 cm² culture flask of human fibroblasts, grown to confluency
50 ml Dulbecco's Modified Eagle Medium (DMEM) supplemented with 10% fetal bovine
    serum and an antibiotic/antimycotic (DMEM + +)
50 ml calcium and magnesium free phosphate buffered saline (CMF-PBS).
10 ml trypsin/EDTA solution (diluted to 1X in CMF-PBS)
Culture hood (optional, but recommended)
37° C humidified CO_2 incubator, set for 10% CO_2
Microscope
Sterile 5 and 10 ml plastic disposable serological pipettes
Sterile 25 cm² tissue culture flasks (Corning)
Sterile tissue culture treated plastic petri dish (Corning)
12x75 mm plastic snap top culture tubes
Pasteur pipettes
Hemocytometer
70% ethanol in squirt bottles
Absolute methanol
Giemsa stain solution (Sigma GS-500)
Trypan blue stain (0.4% in water)

Note to Instructors

The culture medium (DMEM + +) is prepared aseptically by adding 10 ml of fetal bovine serum (Sigma F4884), 1 ml of a 100X antibiotic/antimycotic agent (Sigma A9909) and 1 ml of L-glutamine (100X, Sigma G7513) per 100 ml of medium (check the formulation of the medium to see if addition of L-glutamine is necessary). The trypsin/EDTA solution can be purchased from Sigma (10X, Sigma T9395) and diluted to 1X in CMF-PBS just before use. Human fibroblasts can be purchased from the American Type Culture Collection (ATCC) if they cannot be obtained from someone in your department (follow the instructions from ATCC for breaking out the culture). Once you have fibroblasts in culture, be sure to start subculturing them far enough in advance of the lab so that each student pair will get one flask.

Aliquot the culture medium, CMF-PBS, and trypsin/EDTA solutions into bottles that have been sterilized in an autoclave. Give each student group one bottle of each, which should then be labelled and used by that group, exclusively, for the duration of the exercise. After each use, store the bottles at 4° C in a refrigerator that is accessible to the students. They will then be able to come in and perform their subcultures independently.

One of the biggest problems associated with performing cell culture in a teaching laboratory is contamination. The antibiotic/antimycotic solution added to the culture medium should help to contain gross contamination of the cultures by microorganisms.

In addition, students should be given a thorough demonstration of aseptic technique. A culture hood will also help to reduce contamination. If a hood is not available, an area of the lab should be set aside for use as a "sterile bench". It should be in a remote region of the lab, removed from doors, windows, and the majority of student traffic. This area should be wiped clean with 70% ethanol both before and after each use. Incubators should also be periodically cleaned with 70% ethanol, and a biocide added to the water reservoir will help prevent fungal growth. If cultures or reagents do become contaminated, they should be thrown away immediately to prevent the spread of contamination.

If a CO_2 incubator is not available, the culture vessels can be purged with premixed 10% CO_2 in air and incubated in a regular incubator with the caps tightly closed. Alternately, the use of CO_2 can be circumvented entirely by using a culture medium prepared with HEPES buffer (for example, L-15 Medium Leibovitz), which does not rely on a bicarbonate-carbonic acid buffering system, as do most media.

Procedure

A. Preparation

Obtain a bottle of serile medium, CMF-PBS, and trypsin/EDTA from your instructor. Label the bottles with your initials and the date. Only use these reagents when you subculture the cells. Between uses, the media and other reagents should be kept refrigerated, to help prevent the growth of microorganisms. If your solutions become contaminated (change color and/or become cloudy), report the problem immediately to your instructor and then throw the contaminated solutions out.

In addition to the culture reagents, you will be provided with a tissue culture flask containing human fibroblasts. These cells grow to confluency and must be subcultured every 3 to 4 days. Therefore, you will need to return to the laboratory and subculture your cells at some point in between scheduled lab periods. The cultures will also have to be checked on a daily basis (see step d, Eyeballing Cell Cultures).

Below is a general subculturing outline that can be used for scheduling purposes:

Day 1

Subculture, as demonstrated, the supplied fibroblast culture (see step c, Subculturing Anchorage Dependent Cells). The cells are to be subcultured (or "split") into two 25 cm² culture flasks (a 1:2 split).

Day 4

Subculture *one* flask from Day 1 into (a) two flasks, and (b) one tissue culture treated petri dish (a 1:3 split).

Day 8

(a) Trypsinize *one* flask from Day 4 to obtain a cell suspension (see step f, Counting the Cells). Record the viability and the concentration of the culture. (b) Stain the cells in the petri dish with Giemsa stain (see step e, Giemsa Staining) and examine them using a bright-field microscope. (c) "Feed" the cells in the *second* flask from Day 4, by aseptically pouring off the old medium and adding 5 ml of fresh DMEM + +.

Day 11

Subculture the flask from part (c) Day 8 into two flasks (1:2 split). These cells will be used in the experiment described in chapter 12.

At the end of the second week, you will be asked to assess the overall health of your culture by evaluating the following:

1. The microscopic appearance of the culture each time it was monitored (see Eye-balling Cell Cultures).
2. The final viability (see Counting the Cells).
3. The final concentration of the cells (in cells/ml—see Counting the Cells).
4. A description (or drawing) of the Giemsa stained cells (see Giemsa Staining).
5. A discussion of any problems (contamination, difficulty with the subculturing procedure, etc.).

Before beginning the experiment, read through all of the procedures on the following pages. Then write up a complete schedule. Remember that your cultures must be checked daily, and subcultured every 4 days. The key to success in this process is to be organized and careful with your supplies during the subculture period.

B. Aspetic Technique

Aspetic or sterile technique is the execution of tissue culture procedures without introducing contaminating microorganisms from the environment. Most of the problems related to cell culture are associated with the lack of good aseptic technique. Microorganisms that may contaminate cultures exist everywhere, on the surface of all objects and in the air. A conscious effort must be made to keep them out of a sterile environment.

Newcomers may find that cell culture manipulations are awkward and difficult to master at first. The addition of an antibiotic to the culture media may help to reduce the problems of gross contamination which result from poor sterile technique. Autoclaving renders pipettes, glassware, and solutions sterile, but cannot be used to sterilize culture medium because nutrients in the media may be destroyed by the heat treatment (media must be filter-sterilized).

Following are some general rules to follow to keep your medium, cultures, reagents, and glassware free of contamination:

1. Whenever possible, work in a cell culture hood. This will prevent airborne organisms from entering your cultures.
2. Work areas should be wiped clean with 70% ethanol before and after use. It may help to rinse your hands with ethanol as well or to wear gloves.
3. Never leave sterile flasks, bottles, or petri dishes, open to the environment. Do not remove the cover until the instant you are ready to use it, and replace the cap as soon as you are finished.
4. Never place caps removed from bottles or flasks down on the bench surface. Instead, grasp it in the pinky finger of one hand, and keep it facing down. Once a bottle is open, keep it tilted so that airborne microorganisms will not fall directly into the medium.
5. Sterile pipettes should not be removed from their container until just before use. They should be kept at the sterile bench or in the culture hood only.

6. Do not pipette by mouth. Use a pipette bulb or automatic pipettor when working with solutions. Even though pipettes are plugged with cotton, mycoplasmas from your mouth can still squeeze through the cotton and contaminate your cultures.

7. Use a separate sterile pipette for each manipulation. Do not draw from a bottle more than once with the same pipette.

8. Do not talk, sing, chew gum, or breathe directly on your culture when performing sterile technique.

9. Techniques should be performed as rapidly as possible to minimize exposure to microorganisms.

It is important to note that you must constantly be aware that microorganisms are everywhere (your skin, hands, hair and clothes, for example) and take the proper steps to keep them out of your cultures. When first developing sterile technique you must always be thinking about sterility. Eventually it will all become second nature. Mastering good sterile technique will save you considerable frustration in the future.

C. Subculturing Anchorage Dependent Cells
   (Use sterile technique throughout)
1. Decant the culture medium from the culture flask.
2. Wash the monolayer of cells with calcium and magnesium free phosphate buffered saline (CMF-PBS) by adding 5 ml to the flask, rocking it back and forth gently once or twice, and then pouring the wash solution out of the flask into a container in the cell culture hood.
3. Add 1 ml of 1X trypsin/EDTA solution to the culture flask. Gently rock it back and forth several times to completely cover the growth surface with solution. Set the flask on a level surface, and incubate at room temperature for 10 minutes. This treatment breaks the attachments between the cells and the growth surface, and causes them to "ball up", taking on a rounded appearance.
4. After 10 minutes, add 10 ml of DMEM++ to the flask. The serum in the medium contains trypsin inhibitors that will halt the trypsinization reaction.
5. Gently aspirate the medium up and down in the pipette over the monolayer, to resuspend the cells. The medium will begin to take on a cloudy appearance, as the cells detach and begin to float free.
6. Transfer half of the culture to each of two sterile 25 cm² culture flasks. This is referred to as a "passage" or "split". In this case, the cells in one flask were split into two new vessels—a 1:2 split. Label the flasks with your initials, the date, and the passage number (how many times the cells have been subcultured after the primary culture was made).
7. Incubate the cultures at 37° C with 10% CO_2 in a humidified incubator. The caps on the flasks should not be fully tightened, to allow circulation of gasses.
8. Monitor the cultures on a daily basis (see Eyeballing Cell Cultures). Before you perform any manipulations with the culture flask, be sure to completely tighten the cap. It should be loosened again after it is placed in the incubator.

   In many laboratories, cell counts are performed each time the cells are subcultured. Cells are then "seeded" into culture vessels at specific concentrations, to maximize cell growth. Human fibroblasts grow quickly and usually achieve confluency in 2 to 3 days. When confluency is reached in a 25 cm² flask, it can be assumed that the concentration of cells in the flask is approximately 1×10^7 cells/ml.

90

D. Eyeballing Cell Cultures

The overall health of cells in culture should be assessed on a daily basis and before performing any kind of manipulation in the culture. This can be done quickly and qualitatively by making the following observations:

1. Check the pH of the culture medium by looking at the color of the indicator in the culture medium, phenol red. As a culture becomes more acid, the indicator shifts from red to orange to yellow. In the case of bacterial or yeast contamination, the medium will become quite yellow due to the buildup of acidic microbial metabolic waste products in the medium. As the medium becomes more alkaline, the color shifts from red to a deep purple-red. This is usually related to the buffering system in the medium, and indicates a problem with the CO_2 concentration. As a generalization, cells can tolerate light acidity better than they can tolerate shifts in pH above pH 7.5. Use uninoculated culture medium as a standard to check the pH of your cultures.

2. Using a microscope, check the attachment of the cells to the growth surface. Most of the cells should appear well attached and spread out in a healthy culture (figure 11.3). Floating, round cells can indicate that the cells are dividing; however, they can also mean that the cells are dying. If the majority of the cells in a culture flask are floating, it is likely that the culture is dead.

3. The growth of a culture can be estimated by microscopically following the development of a full cell sheet (confluency). By comparing the amount of surface covered by cells with unoccupied space, you can estimate percentage of confluency. For example, if half of the growth surface appears to be covered with cells, the culture is 50% confluent.

4. Cell shape is an important guide. Round cells in an uncrowded culture are not good (unless they happen to be dividing cells—look for doublets of dividing cells). So called "giant" cells are also an indication that the culture is not doing well. The number of giant cells will increase as a culture ages or declines in well-being.

5. One of the most valuable early guides in evaluating the success of a subculture is the rate at which the cells in a newly established culture attach and spread out. Attachment within an hour suggests that the cells have not been overly traumatized and the culture will grow. Longer attachment times are suggestive, but not a definite sign, of problems.

Figure 11.3. A normal fibroblast in culture. © Keith Roberts and James Barnett

E. Giemsa Staining

1. Pour the culture medium out of the petri dish. The cells, if healthy, will remain attached to the bottom of the dish. Before you begin the staining process, examine the dish microscopically to make sure that there are attached cells present.
2. Rinse the monolayer once with 5 ml of PBS.
3. Fix the cells to the petri dish by adding enough absolute methanol to completely cover the cells. Leave the cells in methanol for at least 10 minutes, then pour it off and allow the dish to completely air dry.
4. Add Giemsa stain solution to the petri dish to completely cover the cells. Stain the cells for 10 minutes. To ensure uniform staining, gently swirl the petri dish every 2 to 3 minutes.
5. Pour off the stain and gently rinse the petri dish in running water. Allow it to completely air dry. Examine the stained cells by placing the petri dish on the stage of a light microscope, and view at 40X and 100X (oil immersion).

F. Counting the Cells

To determine the concentration of cells in a culture and the number of viable (living) cells, the dye exclusion technique is used. Live cells will exclude the dye trypan blue and appear bright and refractile when viewed microscopically while dead or membrane damaged cells take up the dye and appear dark. Cells are counted on a hemocytometer, which can be used to determine the concentration in cells/ml.

1. Prepare a cell suspension in CMF-PBS from the culture in one flask, as follows:
   a. Pour the medium from the culture flask, and wash the monolayer once with 5 ml of CMF-PBS.
   b. Add 1 ml of trypsin/EDTA solution to the monolayer, rocking the flask back and forth several times. Incubate the flask at room temperature for 10 minutes.
   c. Add 10 ml of CMF-PBS to the flask and gently resuspend the cells by pipetting up and down. Pour the suspension into a sterile 15 ml culture tube.
   d. Centrifuge the tube in a clinical centrifuge to pellet the cells (approximately 400 × g). Aspirate and discard the supernatant.
   e. Gently resuspend the cells in 5 ml of CMF-PBS and place the tube on ice.
2. Remove 0.2 ml of the cell suspension, and place it into a 12x75 mm culture tube. Add 0.3 ml of CMF-PBS to the cells, and gently mix.
3. Add 0.5 ml of trypan blue solution to the tube and thoroughly mix the contents by vigorously pipetting up and down with a pasteur pipette (at this point the cells have been diluted 1:5, for a dilution factor of 5).
4. With a coverslip in place, use a pasteur pipette to transfer a small amount of trypan blue/cell suspension to both chambers of a hemocytometer by carefully touching the edge of the coverslip with the pipette tip and allow capillary action to fill the counting chambers. Do not overfill or underfill the chambers.

5. Starting with one of the counting chambers, count all of the cells (living and dead) in the center square and four corner squares of the hemocytometer, as shown in figure 11.4. Keep a separate count of viable (clear and bright) and nonviable (dark) cells.
6. Repeat step 5 on the second counting chamber.
7. If you observe fewer than 200 or greater than 500 cells (20–50 cells per square), repeat the procedure and adjust to an appropriate dilution factor.

DIAGRAM II
STANDARD HEMOCYTOMETER CHAMBER

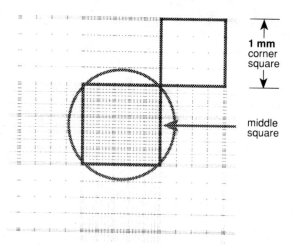

1 mm corner square

middle square

The circle indicates the approximate area covered at 100x microscope magnification (10x ocular and 10x objective). Include cells on top and left touching middle line (○). Do not count cells touching middle line at bottom and right (∅). Count 4 corner squares and middle square in both chambers (one chamber represented here).

DIAGRAM III
CORNER SQUARE (ENLARGEMENT)

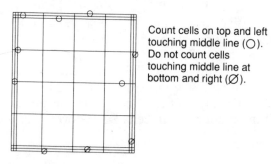

Count cells on top and left touching middle line (○). Do not count cells touching middle line at bottom and right (∅).

Figure 11.4. Using a hemocytometer to perform cell counts. From Sigma Cell Culture Reagents 1990 Catalog. Copyright Sigma Chemical Co., St. Louis, MO.

8. Each square of the hemocytometer represents a total volume of 0.1 mm³ (10^{-4} cm³) under the coverslip. Since one cubic cm is approximately 1 ml, you can determine the cell concentration and the total number of cells using the following calculations:

$$\text{Cells/ml} = \frac{\text{total number of cells}}{\text{number of squares}} \times \text{dilution factor} \times 10^4$$

For example, if 200 cells were counted in 10 squares, the concentration would be:

$$\frac{200}{10} \times 5 \times 10^4 = 1 \times 10^6 \text{ cells/ml}$$

To determine the total cells in the culture:

Total cells = cells/ml × original volume of cell suspension

For example, if the cells were resuspended in 5 ml of solution after trypsinization, the total cell count is:

$$1 \times 10^6 \text{ cells/ml} \times 5 \text{ ml} = 5 \times 10^6 \text{ total cells}$$

You can determine the percentage of viability by dividing the number of viable cells by the total number of cells counted and multiplying by 100. For example, if 180 viable cells were counted in a total of 200 cells, the percentage of viability would be:

$$\frac{180}{200} \times 100 = 90\% \text{ viability}$$

NOTES AND CALCULATIONS

Chapter 12

Cell Adhesion in Human Fibroblasts

Introduction

The extracellular matrix of vertebrate and invertebrate organisms is a complex meshwork of macromolecules found in the extracellular space of most tissues. Until recently it was believed that the matrix was merely an inert component of tissues that served simply as a universal biological "glue"—holding cells together and providing tissues with rigidity. Recent evidence, however, indicates that the extracellular matrix plays an important role in many other biological processes, such as development, cellular proliferation and metabolism.

The macromolecular components of the extracellular matrix, proteins and polysaccharides, are secreted by local cells, especially the fibroblasts. These matrix-forming cells are found distributed throughout the organized meshwork created by interactions between the macromolecules. Together, the matrix and cells constitute the connective tissues, such as cartilage, tendon, and bone (figure 12.1). The amount of connective tissue found in an organ is directed in part by the matrix secreting cells, and varies tremendously. For example, connective tissue is a major component of bones but there is very little in the brain. The macromolecular components of the extracellular matrix and the way that they are organized within a particular tissue is highly variant as well.

The collagens, which are fiber-forming proteins, and the polysaccharide glycosaminoglycans (usually found in the form of proteoglycans) are the two major classes of matrix macromolecules. Glycosaminoglycan molecules interact with the proteoglycans to form a network, with collagen fibers interwoven amongst the strands. The result is a porous yet resilient network of fibers (figure 12.2) that form highly specialized three-dimensional structures.

Elastic tissues such as skin or blood vessels also contain molecules of the rubberlike protein elastin. The elastin molecules interact with one another to form sheets of elastic fibers found interwound with the stiff collagen fibers in the extracellular matrix. This weave of protein fibers allows the tissue to stretch and then recoil in a highly elastic manner yet retain its overall structure.

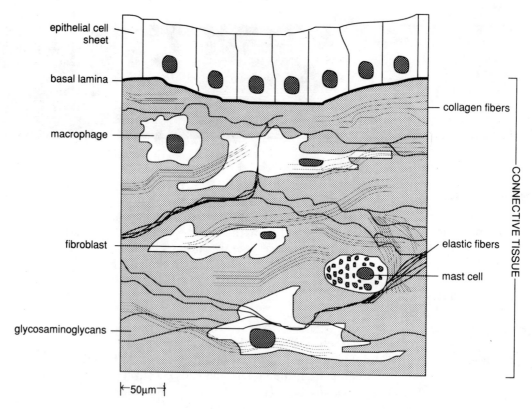

Figure 12.1. A schematic cross section of tissue showing the connective tissue underlying an epithelial cell sheet. From B. Alberts, et al., *Molecular Biology of the Cell*. Copyright © Garland Publishing Inc., New York, NY. Reprinted by permission.

The glycoproteins fibronectin and laminin have also been found in the extracellular matrix. Laminin is a major component of the basal laminae, found just under the outer layers of tissues. The basal laminae are believed to play an important role in the regeneration of injured tissues. Fibronectin is a fiber-forming glycoprotein found in connective tissues. It has also been found on the surface of normal fibroblasts growing in culture.

Transformed fibroblasts removed from tumors and grown in culture were found to have much less fibronectin on their surface than their normal counterparts. When these transformed cells were reinjected into laboratory animals, they rapidly spread throughout body tissues (metastasized) and readily formed tumors. The invasiveness of the transformed fibroblasts could be directly related to the decreased amounts of cell surface fibronectin.

Transformed fibroblasts grown in culture are abnormal in many ways—they do not grow attached to a solid surface and do not spread out but instead appear rounded. They also grow to a very high density, a characteristic related to the loss of contact inhibition

Figure 12.2. The extracellular matrix consists of a network of glycosaminoglycan, proteoglycan and collagen molecules, as shown in this electron micrograph of chick embryo cornea tissue. Robert Trelstad

by the cells. However, if fibronectin is added to these cultures, the cells begin to flatten out and adhere to the growth surface in a manner characteristic of normal cells (figure 12.3). In addition, purified fibronectin has been shown to promote cellular adhesion to collagen, to surfaces, and to other cells, in many different types of cells. Fibronectin may play a role in cell migration in tissues via its influence on this process.

In this experiment, you will study cell adhesion in human fibroblasts using a modified version of a protein "dot-blot" assay. You will dot several different protein solutions onto a nitrocellulose membrane, which has a high affinity for protein. You will add fibroblasts grown in culture to the membrane. When the fibroblasts come in contact with a protein involved in the adhesion process, they will attach to the dot and begin to flatten out. You will fix and stain the membrane, revealing the proteins that act to mediate cell adhesion.

Materials

Nitrocellulose membrane circles (Millipore—HAHY 085–50)
Petri dishes
Micropipettor and tips
Sterile 15 ml culture tubes
Sterile 5 and 10 ml serological pipettes

98

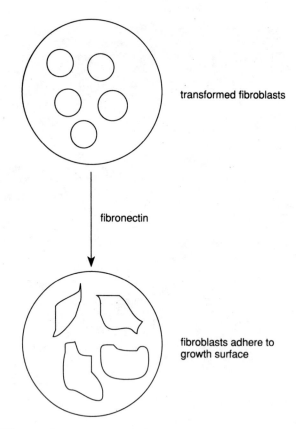

transformed fibroblasts

fibronectin

fibroblasts adhere to
growth surface

Figure 12.3. Fibronectin appears to play an important role in cellular adhesion.

37° C incubator (with 5–10% CO_2, if available)
Clinical centrifuge
Bright-field or phase-contrast microscope
DMEM++
DMEM (serum and antibiotic free)
Calcium and magnesium free PBS (CMF-PBS)
PBS
0.5% Tween 20 in PBS
Trypsin/EDTA solution
Human serum
Conditioned culture medium (DMEM++ removed directly from cultured cells)
Albumin (2 mg/ml in PBS)
Collagen (2 mg/ml in PBS)
Fibronectin (2 mg/ml in PBS)
Culture flask containing fibroblasts grown to confluency (from the previous experiment)
Destain solution (90% methanol, 2% glacial acetic acid)
Amido black stain (45% methanol, 10% acetic acid, 0.1% amido black)

Note to Instructors

Each student group needs one 25 cm² flask of fibroblasts grown to confluency. It is advisable to grow extra flasks of cells, in case there are not enough student-grown cultures from the previous experiment. If there were not any contamination problems, the students' trypsin/EDTA solution and CMF-PBS can be recycled from the previous experiment for use in this experiment as well.

Cells in culture secrete proteins and metabolic waste products into the medium as they grow. Medium that has had cells growing in it is referred to as "conditioned". Before students trypsinize their cultures to prepare the cell suspension, they should remove a small aliquot of conditioned medium (50 μl) and place it on ice to be spotted onto the membrane with the other proteins. All of the protein solutions should be kept on ice while in use.

Procedure

1. Using a soft pencil with a rounded point, draw a grid on a nitrocellulose circle. Number the blocks 1–8, as shown below:
2. Using a micropipettor, carefully dispense 5 μl of each of the protein solutions listed below into a separate block in the grid. Gently touch the tip of the micropipettor to the surface of the membrane, and slowly depress the plunger so that the solution stays localized in one small spot.

Protein Solutions:
1. Human serum
2. DMEM++
3. Conditioned medium
4. Serum free DMEM
5. Albumin
6. Fibronectin
7. Collagen
8. PBS

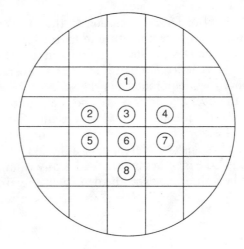

3. Allow the spots to completely air dry, then place the membrane on a piece of filter paper in a 37° C incubator and allow it to incubate for 20 minutes.
4. Add 20 ml of CMF-PBS to a petri dish. Slowly immerse the membrane in the saline solution at a 45° angle to ensure complete wetting of the membrane. Allow the membrane to soak for 5 minutes.
5. Pour the buffer off the membrane in the petri dish. Add 20 ml of 0.5% Tween 20 in PBS (to block nonspecific binding of the fibroblasts to the membrane). Incubate the membrane 20 minutes at room temperature with gentle agitation.

6. While the membrane is soaking in the blocking solution, prepare a fibroblast suspension for use in the cell adhesion assay, as follows:
   a. Pour the medium from the culture flask, and wash the monolayer once with 5 ml of CMF-PBS.
   b. Add 1 ml of trypsin/EDTA solution to the monolayer, rocking the flask back and forth several times. Incubate the flask at room temperature for 10 minutes.
   c. Add 10 ml of CMF-PBS to the flask, and gently resuspend the cells. Pour the suspension into a 15 ml sterile culture tube.
   d. Centrifuge the tube in a clinical centrifuge to pellet the cells (approximately 1000 rpm). Aspirate and discard the supernatant.
   e. Add 5 ml of serum free DMEM to the pellet, and gently resuspend the cells by pipetting the pellet up and down several times.
   f. Centrifuge the tube to pellet the cells, then aspirate and discard the supernatant.
   g. Resuspend the pellet in 15 ml of serum free DMEM and place the tube on ice.
7. Remove the membrane from the blocking solution, and place it into a clean petri dish. Add 20 ml of PBS, and wash for 5 minutes at room temperature with gentle agitation. Pour off the solution, and repeat the wash (for a total of 2 washes in PBS). Pour off and discard the final wash.
8. Add the fibroblast suspension to the membrane in the petri dish. Incubate for 45 minutes at 37° C (with 5–10% CO_2, if available).
9. Pour the cell suspension from the membrane, and wash it 2 times with 20 ml of PBS, as described in step #7.
10. Add 20 ml of Destain solution to the membrane, and wash the membrane for 5 minutes at room temperature to fix the cells on the membrane.
11. Pour off the Destain solution, and add 20 ml of Amido black stain solution. Stain the membrane for 5 minutes at room temperature.
12. Remove the membrane from the stain, and rinse it 3 or 4 times in Destain solution until the background is white and the stained cells are visible (as blue-black dots).
13. Examine the membrane macroscopically, and note the location of the stained cells. Look also at the intensity of the color obtained, which provides a rough indication of the fibroblasts' preference for adhesion.
14. Examine the membrane microscopically (illumination must come from above), and note the appearance and the density of the attached fibroblasts.

NOTES AND CALCULATIONS

Appendix I

Chemical Formulations

Trypan blue stain

0.4% (w/v) trypan blue in distilled water. Store at room temperature.

0.7% saline

0.7% (w/v) NaCl in distilled water. Autoclave. Store at room temperature.

Orcein stain

orcein	2 g
glacial acetic acid	45 ml
distilled water	55 ml

Stir (with heat) for 15 minutes to dissolve orcein. Cool solution to room temperature, then filter stain through Whatman #1 filter paper. Store in a brown bottle at 4° C.

0.075 M KCl

KCl	0.56 g
distilled water	to 100 ml

Autoclave. Store at room temperature.

0.025 M potassium phosphate buffer, pH 6.8

KH_2PO_4	0.17 g
K_2HPO_4	0.21 g
distilled water	to 100 ml

Autoclave. Store at room temperature.

1 % Trypsin

1% (w/v) in phosphate buffered saline (PBS). Store at 4° C.

1 N HCl

concentrated HCl	2 ml
distilled water	22 ml

Store at room temperature.

1 N KOH

KOH	5.6 g
distilled water	to 100 ml

Store at room temperature in a plastic bottle.

1 N NaOH

NaOH	4 g
distilled water	to 100 ml

Store at room temperature in a plastic bottle.

1 M Tris-HCl pH 8.0

Tris base	5.3 g
Tris HCl	8.9 g
distilled water	to100 ml

Autoclave. Store at room temperature.

500 mM EDTA

disodium EDTA-2H$_2$O	18.6 g
distilled water	50 ml

Stir to dissolve EDTA. Adjust pH to 8.0 with 1 N NaOH. Bring volume to 100 ml with distilled water and autoclave. Store at room temperature.

10% SDS

10% SDS (w/v) in distilled water. Store at room temperature.

TE pH 8.0 (10 mM Tris-HCl pH 8.0, 1 mM EDTA)

1 M Tris-HCl pH 8.0	1 ml
500 mM EDTA	0.2 ml
distilled water	to 100 ml

Autoclave. Store at room temperature.

Mini-prep Solution I

1 M Tris-Cl pH 8.0	2.5 ml
500 mM EDTA	2 ml
distilled water	to 100 ml

Autoclave. Store at 4° C. Just before use, add 1.25 ml of 20% glucose (w/v in distilled water) and 20 mg of lysozyme, to 5 ml of Solution I.

Mini-prep Solution II

1 N NaOH	1 ml
10% SDS	0.5 ml
distilled water	3.5 ml

Make just prior to use. Keep at room temperature.

Mini-prep Solution III

5 M potassium acetate	60 ml
glacial acetic acid	11.5 ml
distilled water	28.5 ml

Store at room temperature.

TY broth + ampicillin

tryptone	1 g
yeast extract	0.5 g
NaCl	1 g
1 N NaOH	0.2 ml
distilled water	to 100 ml

Autoclave at 121° C for 15 mimutes. After broth has cooled, add 5 mg (or 0.1 ml of a 50 mg/ml stock solution) of ampicillin and gently swirl. Store medium at 4° C.

1 % Bromophenol blue

bromophenol blue	0.1 g
distilled water	10 ml

Dissolve with gentle stirring. Store at room temperature.

1% Xylene Cyanol

xylene cyanol	0.1 g
distilled water	10 ml

Dissolve with gentle stirring. Store at room temperature.

6X Sample Buffer (for agarose gel electrophoresis)

1% bromophenol blue	2.5 ml
1% xylene cyanol	2.5 ml
glycerol	3 ml
distilled water	2 ml

Store at room temperature.

50X TAE

Tris base	24.2 g
glacial acetic acid	5.7 ml
500 mM EDTA	10 ml
distilled water	to 100 ml

Do not autoclave. Store at room temperature.

Ethidium bromide stain

ethidium bromide	100 mg
distilled water	10 ml

Wrap container in aluminum foil. Store at 4° C.

Phosphate Solution A (0.2 M NaH_2PO_4)

NaH_2PO_4-H_2O	27.6 g
distilled water	to 1 liter

Autoclave. Store at room temperature.

Phosphate Solution B (0.2 M Na_2HPO_4)

Na_2HPO_4-$7H_2O$	53.65 g
distilled water	to 1 liter

Autoclave. Store at room temperature.

0.01 M phosphate buffer, pH 7.2

Phosphate Solution A	28 ml
Phosphate Solution B	72 ml
distilled water	to 2 liters

Store at 4° C.

1 M phosphate buffer, pH 7.2

NaH_2PO_4-H_2O	38.6 g
Na_2HPO_4-$7H_2O$	193.0 g
distilled water	to 1 liter

Store at 4° C.

See Appendix III for preparation of polyacrylamide gels for SDS-PAGE.

0.5 M Tris-HCl pH 6.8

Tris base	6 g
distilled water	60 ml

Adjust pH to 6.8 with 1 N HCl. Bring volume to 100 ml. Autoclave and store at room temperature.

5X SDS-PAGE Running Buffer

Tris base	1.5 g
glycine	7.2 g
SDS	0.5 g
distilled water	to 100 ml

Store at room temperature. Dilute to 1X before use.

SDS-PAGE Sample Buffer

distilled water	4 ml
0.5 M Tris-HCl pH 6.8	1 ml
glycerol	0.8 ml
10% SDS	1.6 ml
β-mercaptoethanol	0.4 ml
1% bromophenol blue	0.2 ml

Aliquot into 0.5 ml fractions. Store at $-20°$ C.

Coomassie blue stain

Coomassie blue R250	2.5 g
methanol	400 ml
glacial acetic acid	100 ml
distilled water	500 ml

Store at room temperature. Use stain solution within two weeks.

SDS-PAGE Destain I

methanol	250 ml
glacial acetic acid	50 ml
distilled water	200 ml

Make just prior to use.

SDS-PAGE Destain II

glacial acetic acid	35 ml
methanol	25 ml
distilled water	440 ml

Make just prior to use.

Mitochondria Isolation Buffer

D-mannitol	27.3 g
KH_2PO_4	0.4 g
K_2HPO_4	1.2 g
distilled water	400 ml

Adjust pH to 7.2 with 1 N KOH. Bring volume to 500 ml with distilled water. Store at 4° C.

Mitochondria Assay Buffer

D-mannitol	27.3 g
KH_2PO_4	0.4 g
K_2HPO_4	1.2 g
KCl	0.4 g
$MgCl_2 \cdot 6H_2O$	0.5 g
distilled water	400 ml

Adjust pH to 7.2 with 1 N KOH. Bring volume to 500 ml with distilled water. Store at 4° C.

0.4 M sodium azide

sodium azide	0.26 g
distilled water	to 100 ml

Sodium azide is a poison. Store at room temperature.

5×10^{-4} M DCIP

2,6-dichlorophenolindophenol (sodium salt)	14.5 mg
distilled water	to 100 ml

Make just before use.

0.2 M succinate

sodium succinate—$6H_2O$	5.4 g
distilled water	90 ml

Adjust pH to 7 with 1 N HCl. Bring volume to 100 ml with distilled water. Store at 4° C.

Amido black stain

amido black	0.1 g
methanol	45 ml
glacial acetic acid	10 ml
distilled water	45 ml

Make just prior to use.

Amido black destain

methanol	90 ml
glacial acetic acid	2 ml
distilled water	8 ml

Make just prior to use.

Appendix II

Equipment and Supplies

Equipment

Electrophoresis equipment (SDS-PAGE, AGE, power supply)	Bio-Rad Hoefer Schleicher and Schuell
Spectrophotometer (Spectronic 20)	Milton Roy
Microfuge	VWR Fisher
Micropipettors	Fisher Bio-Rad
Gel drying frame (air)	Hoefer Research Products Inc.
Photography and autoradiography equipment and supplies	Kodak Sigma

Supplies

Chemicals and reagents	Amresco Sigma Fisher
Dialysis tubing and clamps	Fisher
Microfuge tubes, micropipettor tips	Bio-Rad Fisher
Nitrocellulose	Millipore
Drosophila	Carolina Biological Supply Company
In vitro translation kit	Promega
^{35}S-methionine (*in vitro* translation grade)	New England Nuclear Amersham
PlasmidQuik (to prepare plasmid DNA)	Stratagene
Restriction enzymes (and 10X buffer)	Boehringer Mannheim Biochemicals
Purified λ and pBR322 DNA	Boehringer Mannheim Biochemicals
pRPC245 (in *E. coli* MM294)	available from author (see below)
SDS-PAGE molecular weight standards	Bio-Rad Pharmacia

Human fibroblasts	ATCC.
Tissue culture treated plasticware (culture flasks, tubes, petri dishes)	Fisher (Corning)
Culture media (DMEM, FBS, glutamine, antibiotic/antimycotic, trypsin/EDTA, PBS, CMF-PBS)	Sigma Cell Culture Gibco

The plasmid pRPC245 is a specialized construction which can be obtained from the author. Other plasmids with interesting restriction maps can be easily substituted for pRPC245 in chapter 5.

E. coli MM294 cells transformed with pRPC245 will be sent upon request. The plasmid must then be purified by scaling up a mini-prep procedure (described in chapter 4), or by using a plasmid DNA extraction kit (PlasmidQuik by Stratagene is recommended).

Requests for the plasmid can be sent to:

Holly Ahern
Department of Biological Sciences
State University of New York at Albany
Albany, NY 12222

Suppliers

American Type Culture Collection
Amersham, Inc.
Amresco
Bio-Rad Laboratories
Boehringer Mannheim Biochemicals
Carolina Biological Supply Company
Corning
Fisher Scientific
Gibco
Hoefer Scientific Instruments
Kodak Laboratory and Research Products
Millipore
Milton Roy
New England Nuclear Research Products
Pharmacia LKB Biotechnology
Promega Corporation
Research Products Inc.
Schleicher and Schuell
Sigma Chemical Co.
Stratagene
VWR

Appendix III — Additional Procedures

Subculturing Bacterial Cells from Stab Medium

1. Using sterile technique, dispense 2 ml of TY broth with ampicillin into a sterile glass culture tube (to select for cells carrying the plasmid pRPC245, which carries an ampicillin resistance gene).
2. Sterilize an inoculating loop by holding it in the hottest part of a flame until it glows red.
3. Allow the loop to cool for approximately 5 seconds. Insert the sterilized loop into the stab medium containing the bacterial culture.
4. Withdraw the loop, and transfer the inoculum to the broth medium. Shake the loop gently in the liquid.
5. Grow the cells overnight at 37° C. The broth culture will be cloudy if the subculture was successful.

Subculturing from Broth Medium

Liquid cultures can be subcultured by dispensing the required amount of medium (TY + ampicillin for *E. Coli* MM294/pRPC245) into a large sterile tube or sterile flask (with cap). Transfer bacterial cells to the new medium using an inoculating loop as described above. Incubate the cultures overnight at 37° C. Cultures containing more than 50 ml of medium should be incubated in a shaking water bath at 37° C.

Storage of Bacterial Cultures

Bacterial cells can be stored frozen at −20° C for years in 50% glycerol. Dispense 1 ml of glycerol into cryopreservation vials, and sterilize by autoclaving. Store the sterile glycerol at −20° C. To freeze bacteria cultures, grow up a 1 ml overnight culture (in TY broth plus appropriate antibiotic), and aseptically transfer it to a vial of cold sterile glycerol. Mix the contents of the vial completely, and immediately place at −20° C.

To revive a frozen bacterial culture, aseptically transfer a loopful of the cells in glycerol to 2 ml of growth medium (TY + ampicillin for *E. coli* MM294/pRPC245) in a sterile test tube. Grow the cells for 24 to 48 hours at 37° C.

Pouring Polyacrylamide Gels for SDS-PAGE

Solutions:

1. Monomer Stock Solution (30% acrylamide)

Acrylamide (caution: neurotoxin)	58.4 g
Bis	1.6 g
distilled water	to 200 ml

   Wear a mask when preparing this solution. Store at 4° C in a brown or aluminum foil covered bottle.

2. Resolving Gel Buffer

Tris base	18.15 g
distilled water	80 ml

   Adjust to pH 8.8 with 1 N HCl. Bring the volume to 100 ml distilled water. Autoclave and store at room temperature.

3. Stacking Gel Buffer

Tris base	3.0 g
distilled water	40 ml

   Adjust to pH 6.8 with 1 N HCl. Bring the volume to 50 ml with distilled water. Autoclave and store at room temperature.

4. 10% SDS

SDS	10 g
distilled water	to 100 ml

   Store at room temperature.

5. Initiator

Ammonium persulfate	50 mg
distilled water	0.5 ml

   Make fresh immediately before use.

6. Resolving Gel Overlay

Resolving Gel Buffer (Tris pH 8.8)	25 ml
10% SDS	1 ml
distilled water	to 100 ml

Pouring the resolving gel (12% acrylamide, pH 8.8)

	mini-gel (10 ml)	large gel (30 ml)
Monomer stock solution	4 ml	12 ml
Resolving gel buffer	2.5 ml	7.5 ml
distilled water	3.3 ml	10 ml

Add the reagents to a side arm flask and deaerate.

10% SDS	100 μl	300 μl
10% ammonium persulfate	50 μl	150 μl
TEMED	5 μl	15 μl

Add the ammonium persulfate and TEMED immediately before pouring the gel. Swirl the solution gently, and use a pipette to fill the space between the glass plates in the gel assembly (the glass plate sandwich should be assembled according to manufacturer's instructions). Do not fill the sandwich completely; leave enough space on the top of the gel for a stacking gel (approximately 1 inch below the comb). Carefully overlay the resolving gel with Resolving Gel Overlay.

When the resolving gel has polymerized (a distinct boundary will be visible), pour off the overlay and rinse the surface of the gel once with distilled water. Insert a piece of filter paper between the glass plate to blot all traces of water from the surface of the gel.

Pouring the stacking gel (4% acrylamide, pH 6.8)

	10 ml
Monomer stock solution	1.3 ml
Stacking gel buffer	2.5 ml
Distilled water	6.1 ml

Add the reagents to a side-arm flask and deaerate as before.

10% SDS	100 μl
10% ammonium persulfate	50 μl
TEMED	10 μl

Place the comb between the glass plates. Add the ammonium persulfate and TEMED, and gently swirl the solution to mix. Pipette the stacking gel solution into the sandwich on top of the resolving gel, filling the spaces around the comb.

When the stacking gel has polymerized, remove the comb from the top of the gel sandwich. Rinse the sample wells once with distilled water. Insert the gel into the electrophoresis chamber of the apparatus (according to manufacturer's instructions) and fill the sample wells with Running Buffer. Treated protein samples can be loaded onto the gel at this point.